Synthese Library

Studies in Epistemology, Logic, Methodology, and Philosophy of Science

Volume 420

The aim of *Synthese Library* is to provide a forum for the best current work in the methodology and philosophy of science and in epistemology. A wide variety of different approaches have traditionally been represented in the Library, and every effort is made to maintain this variety, not for its own sake, but because we believe that there are many fruitful and illuminating approaches to the philosophy of science and related disciplines.

Special attention is paid to methodological studies which illustrate the interplay of empirical and philosophical viewpoints and to contributions to the formal (logical, set-theoretical, mathematical, information-theoretical, decision-theoretical, etc.) methodology of empirical sciences. Likewise, the applications of logical methods to epistemology as well as philosophically and methodologically relevant studies in logic are strongly encouraged. The emphasis on logic will be tempered by interest in the psychological, historical, and sociological aspects of science.

Besides monographs *Synthese Library* publishes thematically unified anthologies and edited volumes with a well-defined topical focus inside the aim and scope of the book series. The contributions in the volumes are expected to be focused and structurally organized in accordance with the central theme(s), and should be tied together by an extensive editorial introduction or set of introductions if the volume is divided into parts. An extensive bibliography and index are mandatory.

More information about this series at http://www.springer.com/series/6607

Holger Andreas

Dynamic Tractable Reasoning

A Modular Approach to Belief Revision

Springer

Holger Andreas
Department of Economics, Philosophy,
and Political Science
University of British Columbia (Okanagan)
Kelowna, BC, Canada

Synthese Library
ISBN 978-3-030-36232-4 ISBN 978-3-030-36233-1 (eBook)
https://doi.org/10.1007/978-3-030-36233-1

This Springer imprint is published by the registered company Springer Nature Switzerland AG.
The registered company address is: Gewerbestrasse 11, 6330 Cham, Switzerland

Dedicated to Leonard

Modern logic is undergoing a cognitive turn,
side-stepping Frege's 'anti-psychologism'.

Johan van Benthem [112]

Preface

When working on theoretical terms during my PhD thesis, I became intrigued by the Sneed formalism and its set-theoretic predicates. This formalism struck me as a natural continuation of Carnap's logic of science. But I was not convinced by Stegmüller's arguments against a broadly axiomatic and syntactic approach to scientific theories, so I devised an axiomatic variant of the structuralist representation scheme of scientific knowledge.

When looking for applications of my work on structuralism (and opportunities for further funding), I started to explore connections to cognitive science and knowledge representation. I was excited to observe that set-theoretic concepts of the Sneed formalism may help us translate Minsky's ideas about frames into the language of logic and set theory. Thus, it seemed possible to reintegrate frames into the logic-oriented approach to human cognition and artificial intelligence. Hence, I tried to carry on Minsky's research programme while being unconvinced by his arguments against the logic-oriented approach. This endeavour resulted in a novel logic, called *frame logic*.

For Minsky, a major motivation to explore frames was to explain the effectiveness of common-sense thought. How on earth is it possible that human minds process information fast and reliably? According to the present state of the art in computational complexity theory and belief revision, this is not possible at all. Rational belief revision is not tractable. That is, it is not computationally feasible to change one's beliefs – when receiving new epistemic input – in such a manner that certain rationality postulates and logical constraints are respected. So, Minsky seemed to be right, after all, in claiming that logic-oriented approaches to human cognition are on the wrong track. I found this hard to accept, despite all theoretical results pointing towards this conclusion. I therefore explored means to achieve tractability using frame concepts and ideas about truth maintenance. This endeavour resulted in a truth maintenance system for a structuralist theory of belief changes, which in turn leads to a novel belief revision scheme. The belief revision scheme is, in fact, tractable for frame logic.

The present book, divided in two parts, develops frame logic and the truth maintenance-inspired belief revision scheme in a stepwise fashion. Part I starts with an exposition of the general problem of dynamic tractable reasoning, and a chapter on frames follows. Then, the reader is introduced to belief revision theory and defeasible reasoning. Part II starts with an axiomatic, Carnapian account of structuralist theory representation. This account is merged with a specific belief revision scheme, resulting in a structuralist theory of belief changes. Thereupon, a truth maintenance system is devised. Finally, frame logic and a novel belief revision scheme will be developed on the basis of the truth maintenance system.

This book brings together different frameworks from different scientific disciplines, so there are a number of people I would like to acknowledge. First of all, I am very grateful to C. Ulises Moulines for his open-mindedness towards my ideas about a syntactic approach to structuralist theory representation when he co-supervised my PhD thesis. My first encounter with belief revision and frames dates back to a postdoctoral stay at Stanford. So, I would like to thank Johan van Benthem and Yoav Shoham for an introduction to dynamic epistemic logic and nonmonotonic reasoning. Mark Musen introduced me to frames. Heinrich Herre from the University of Leipzig kindly supported my research immediately after completion of the PhD, and encouraged me to explore connections between structuralism and knowledge representation.

After my research stay at Stanford, I enjoyed working as an assistant professor, first at the Chair of Andreas Bartels in Bonn and second at the Chair of C. Ulises Moulines at LMU Munich. Moreover, I would like to thank Stephan Hartmann and Hannes Leitgeb for integrating me into Munich Center for Mathematical Philosophy; special thanks are due to Stephan Hartmann for supporting my research after completion of the habilitation. The earlier versions of Chaps. 1 to 7 of this book formed the core part of my habilitation at LMU Munich in 2012. I would like to thank all the committee members for their very valuable comments that helped me improve these chapters: Gerhard Brewka, Hannes Leitgeb, C. Ulises Moulines and Hans Rott.

Not too long after the habilitation, I accepted an offer from the University of British Columbia for a tenure track position. Moving to Canada with the family was quite a challenge. Also, it seemed time to take a break from computational complexity theory and interdisciplinary work. I resumed my work on structuralist belief changes after this break and wrote Chap. 8 of the present book, which expounds frame logic and the truth maintenance-inspired belief revision scheme. I would like to express my sincere gratitude to the department heads, Helen Yanacopulos and Andrew Irvine, for their kind support of my research and to my colleagues, Manuela Ungureanu, Giovanni Grandi, Jim Robinson and Dan Ryder, for a very supportive and cordial work environment. Andrew Irvine and Dan Ryder helped with very valuable comments on the book proposal and the introduction of the book. Moreover, special thanks are due to Josef Zagrodney and Mario Günther for proofreading. Of course, I remain responsible for all the mistakes.

The research stay at Stanford was supported by the DAAD (the German Academic Exchange Service). The Centre for Advanced Studies at LMU Munich, which awarded me a fellowship in the fall term of 2011, provided an excellent research environment for completing the habilitation. Before I came to Canada, I was a Heisenberg fellow of the DFG, the German Research Council.

Finally, I want to thank my wife, Sabine, and my children, Leonard, Laurenz and Mathilde, for their invaluable emotional support and understanding. This book is dedicated to Leonard, the oldest one, which implies a commitment to write two more books to be dedicated to the younger ones.

Kelowna, BC, Canada Holger Andreas

Contents

List of Figures

Chapter 1
Introduction

In this book, we aim to lay bare the logical foundations of tractable reasoning. We draw on Marvin Minsky's [87] seminal work on frames, which has been highly influential in computer science and, to a lesser extent, in cognitive science. However, only very few attempts have explored ideas about frames in logic. Such is the intended innovation of the present investigation.

In cognitive science, the idea that the human mind works in a modular fashion continues to be very popular. This idea is also referred to as the *massive modularity hypothesis* [100]. The present logical investigation of modularity is directed toward two unresolved problems that have arisen in cognitive science and have yet to receive a proper solution. First, the problem of tractable reasoning, and second the problem of transmodular reasoning with a modular architecture. The first problem concerns the cognitive feasibility of global inferential reasoning: we know very little about the cognitive and logical means that allow us to draw inferences that involve larger amounts of language and memory. Global inferential reasoning appears to be intractable.

Ideas about a modular architecture of human cognition promise to explain how human minds cope with the challenges of computational complexity [41, 100]. However, cognitive scientists have not delivered a more detailed demonstration that modularity actually helps achieve tractability. This gives rise to the second problem: human reasoning is often transmodular and is therefore not confined to a small set of beliefs in a specific domain. A decomposition of reasoning into modular operations thus is needed to maintain the modularity hypothesis for central cognition.

Particularly challenging from a computational perspective is the operation of changing one's beliefs in light of new epistemic input. This is so for two reasons. First, a new piece of information may affect various parts of our overall body of beliefs. Second, new information, if trustworthy, may force us to retract some of our present beliefs. Our reasoning is sometimes dynamic in the sense that we retract certain premises and conclusions that have been accepted so far. This observation gave rise to new areas in formal epistemology and logic: belief revision

H. Andreas, *Dynamic Tractable Reasoning*, Synthese Library 420,
https://doi.org/10.1007/978-3-030-36233-1_1

and nonmonotonic reasoning. Research in these areas is dedicated to the problem of dynamic reasoning, and we now have a variety of approaches to this type of reasoning. But the problem of dynamic tractable reasoning has remained largely unsolved.

The apparent computational infeasibility of inferential reasoning – be it dynamic or static – is a major objection to the *computational theory of mind* and the *language-of-thought* hypothesis propounded by Fodor [55, 59]. But the problem is not tied to these two theoretical hypotheses: as soon as we acknowledge that human minds engage in drawing inferences, we face the problem of computational and cognitive feasibility. This problem does not result from a particular cognitive paradigm.

Why are belief changes computationally challenging, if not even computationally infeasible? Why do these computational issues matter for our understanding of human cognition? Why should a modular account of human reasoning – in terms of frames and frame concepts – help achieve tractability? In what follows, we shall outline an answer to these questions.

1.1 Apparent Intractability of Belief Changes

In science and everyday life, our beliefs are changing continuously. Belief changes are often initiated by a new piece of information. More specifically, we can distinguish between two types of impact that a new piece of information may have on our present beliefs: first, the new epistemic input may allow us to infer further consequences from our present beliefs. Second, we may be forced to give up some of our present beliefs because the new epistemic input is inconsistent with what we presently believe.

This simplified variant of a story by Gärdenfors [61, p. 1] exemplifies how new epistemic input may force us to give up some of our present beliefs. Oscar used to believe that he had given his wife a ring made of gold at their wedding. Later, he realises that his wedding ring has been stained by sulfuric acid. However, he remembers from high school chemistry that sulfuric acid does not stain gold. As he could not deny that the ring is stained and as he also believed that their wedding rings are made of the same material, his beliefs implied a contradiction. Hence, the new epistemic input – i.e., the observation that a certain liquid (believed to be sulfuric acid) stains his wedding ring – forces Oscar to give up some of his present beliefs. From a logical point of view, there are at least three options to regain consistency: (i) he could retract the belief that sulfuric acid stained the ring, (ii) he could call into question that sulfuric acid does not stain gold, and finally, (iii) he could give up the belief that their wedding rings are made of gold. As he paid a somewhat lower price for their wedding rings than might normally be expected, he found himself forced to accept the third option.

The logical study of belief changes has given rise to a new discipline in philosophical logic called *belief revision theory*. For the most part, this discipline has been founded by Alchourrón, Gärdenfors and Makinson [1]. It is for this reason also referred to as *AGM*, or *AGM-style, belief revision theory*. The original AGM theory consists in an account of the laws and the semantics of belief changes. A large number of further belief revision schemes have been devised in the wake of the AGM account. Each of these comes with a specific analysis of belief changes.

Why should it be computationally and cognitively impossible to perform belief changes in a rational way? In brief, the answer is that the amount of calculation demanded for a single belief change is likely to exceed the cognitive power of our mind. Theoretical arguments showing this come from the theory of computational complexity and related complexity results for standard, logical approaches to belief revision [89]. These results strongly suggest that belief revision is an *intractable* problem. In more technical terms: if $\mathbf{P} \neq \mathbf{NP}$ (which is quite firmly believed by most computer scientists), then the problem of rationally revising one's beliefs is intractable. We shall explain the computational complexity of belief changes in greater detail in the course of this investigation.

Intractability of a problem means, in less technical terms, that our best computers take too much time to solve non-trivial instances of the problem, despite all the progress in the development of hardware that we have witnessed in past decades. To give an example, no efficient method has been found for determining the satisfiability of a propositional formula. For, the number of relevant valuations grows exponentially with the size of such a formula so that, in the worst case, there is an exponential growth of the number of computation steps needed to test satisfiability.

Cherniak [36, p. 93] has given a telling illustration of how devastating this type of exponential growth is for the feasibility of determining whether or not a propositional formula (or a set of such formulas) is satisfiable: suppose our super computer is able to check whether a given propositional valuation verifies the members of a set of propositions in the time that light takes to traverse the diameter of a proton. Assume, furthermore, that our propositional belief system contains just 138 logically independent propositions. If so, the estimated time from the Big Bang to the present would not suffice to go through all valuations of the system of propositions. Hence, even this super computer would not provide us with the means to test consistency of a moderately complex propositional belief system in such a manner that the test is conclusive for any possible input. We might of course be lucky and find a propositional valuation that verifies all items of the belief system after going through only a few valuations, but it is much more probable that we will be unlucky.

Exponential growth of the computation steps needed to solve a problem, in the worst case, is considered a mark of intractability. Tractability of a problem, by contrast, means that it can be solved with a reasonable number of computation steps. To be more precise, a problem is tractable if and only if the number of computation steps needed to solve an arbitrary instance of it is bounded by a polynomial function whose argument is a parameter that characterises the syntactic size of the instance. The computation of arithmetic functions, such as addition, multiplication, etc., for natural and rational numbers is tractable because corresponding algorithms exhibit no exponential growth of the number of computation steps.

The problem of revising a belief system is computationally even harder than the problem of testing satisfiability of a propositional formula. Belief revision must therefore be considered an intractable problem, at the present state of the art. As

Nebel [89, p. 121] puts it, "The general revision problem for propositional logic appears to be hopelessly infeasible from a computational point of view because they are located on the second level of the polynomial hierarchy." It is fair to say that there is no reasonably expressive and tractable belief revision scheme to be found in the computer science literature to date.[1]

Why is the problem of rational belief revision even harder than testing satisfiability of a propositional formula? The runtime of algorithms that determine belief changes exactly is not only exponential, but super-exponential. That is, these algorithms require exponentially many calls of a subroutine that consists of exponentially many computation steps. Few attempts have been made toward approximate solutions to the general problem of belief revision.[2] And the potentialities of a modular approach to tractable reasoning and belief revision have not yet been fully exploited.

1.2 Computational Theory of Mind

The distinction between tractable and intractable problems has a certain bearing on the understanding of human cognition, as observed by Cherniak [36], Fodor [58], Johnson-Laird [74], Stenning and van Lambalgen [107], Woods [120] and others. Human minds are far more creative than computers and for this reason are certainly distinct from the latter. But they are not capable of executing algorithms and computations faster and more reliably than computers. As for belief revision, there is a large class of problems that appear to require formal reasoning and calculation more than creativity.

In cognitive science, the *computational theory of mind* (CTM) is a research programme that aims to exploit presumed commonalities between computers and the biological machinery underlying our cognition. It follows quite directly from CTM that any problem that is intractable for computers is intractable for human minds as well. As we shall see later on, rather weak formulations of CTM suffice to show this implication. In particular, CTM implies that we are not capable of obeying the norms of standard approaches to rational belief revision. We do not even have a clear idea of how we could approximately conform to the norms of rational belief change.

Modularity, on the other hand, appears to be a means to escape the combinatorial explosion of belief formation. This is so, however, only if we can decompose global reasoning into modular operations. For, belief formation is often a global affair in the sense that new pieces of information potentially impact a large variety of beliefs across a number of different domains. In molecular biology, for example, a new finding about the homology of two protein sequences may imply hypotheses

[1] See again Nebel [89], who has given the most comprehensive survey of the complexity results of belief revision schemes. Those belief revision schemes that yield, together with a tractable logic, a tractable determination of belief changes are highly counter-intuitive. We shall be more explicit about this in Sect. 1.8 and Chap. 7.

[2] See [115] and [2]. We shall discuss these proposals in Chap. 7.

about their functional similarity, which in turn may be relevant to our theories about specific biochemical pathways. Even in daily life it happens that new information has an impact across domains. Last winter, I met a passionate skier who sold his property at the local ski hill because of a few mediocre winters in a row and because of scientific evidence for global warming.

We are unable, in general, to tell in advance for which beliefs a new epistemic input is relevant. The modularity hypothesis, *by contrast*, asserts that cognitive modules work in a domain-specific way and are encapsulated. In order to maintain the modularity of central cognition, one therefore must resort to an account of interacting modular units of reasoning [35]. This idea, however, has not yet been formulated in a precise manner. We lack, in particular, something like a logical analysis of modular reasoning. Cognitive scientists have not yet been able to prove that a modular account of human cognition explains how global inferential reasoning is cognitively feasible.[3] At the same time, considerations of tractability have been used to advance the massive modularity hypothesis in cognitive science. The presumed intractability of an a-modular account of human cognition serves as one of three core arguments in favour of this hypothesis [41].

In a similar vein, Fodor [58] has observed an impasse in cognitive science: the undeniable existence of global cognition, i.e., cognition that involves larger amounts of language and knowledge, is fundamentally at odds with the computational theory of mind, given the limited success of logic-oriented artificial intelligence and the difficulties of computing global inferences reliably and fast. Fodor observes that we simply have no idea how global cognition is feasible for minds with finitely bounded resources of computation and memory (see [58, Chaps. 2 and 3] and [59, Chap. 4.4]).

Who is the culprit for this impasse? On the face of it, there appear to be only two candidates: CTM and the view that formal logic has some role to play in an account of human cognition. So, shall we discard CTM or some variant of it? Fodor [58] introduces CTM in the more specific sense that higher cognitive processes are classical computations, i.e., computations that consist of operations upon syntactic items. This understanding of CTM forms the core of the classical computationalist paradigm in cognitive science [20]. Broadly logic-oriented research on human cognition has been driven by this paradigm. Assuming a syntactic nature of computation appears to justify applying the computational complexity theory to human cognition in the first place. So it seems natural to discard CTM – more precisely, the syntactic formulation of CTM – in order to solve our problem. Computational complexity theory, however, pertains to human cognition quite independently of this assumption since the scope of complexity theory is governed by the *Church-Turing thesis*. This thesis asserts that any physically realisable computation device – whether it is based on silicon, neurones, DNA or some other technology – can be simulated by a Turing machine [14, p. 26]. Issues of computational complexity, therefore, pertain to human cognition quite independently of a commitment to a syntactic variant of CTM.

[3] The famous work on bounded rationality by Gigerenzer [64, 65] is not concerned with modularity.

Shall we discard, then, logical approaches to human reasoning and cognition? The *connectionist* paradigm in cognitive science is motivated by the significant success of artificial neural networks in pattern recognition and the relatively minor success of *good old-fashioned artificial intelligence*, which is logic-oriented. Proponents of the connectionist paradigm have suggested that formal logic has hardly any explanatory value for human intelligence [39]. It is undeniable, however, that scientific and quotidian reasoning have a genuine propositional structure since reasoning and argumentation essentially consists in making inferential transitions from antecedently acquired or accepted propositions. It has been moreover shown that propositional reasoning can be implemented by means of neural networks [79]. Stenning and van Lambalgen [107], Johnson-Laird [74] and others have successfully supplemented logical approaches to human cognition with empirical research in psychology.

One might also try to dissolve issues of tractability by emphasising the normative role of logic. Logical systems, one could argue, explicate norms of reasoning without actually describing human reasoning. But even a purely normative view of logic would not solve the problem under consideration. In order for a standard to have a normative role, it must be possible to, at least approximately, meet it. A norm that we cannot obey – neither exactly nor approximately – cannot be considered a norm in the first place. *Ought* implies *Can*.[4] Following van Benthem [112], we view the role of logical systems in an analysis of human reasoning as partly normative and partly descriptive.

A closer look at our problem reveals more culprits to consider. There might be something wrong with the distinction between tractable and intractable problems in the theory of computational complexity. In particular, the decision to focus on worst-case scenarios in the original distinction is open to question.[5] This line, however, is not pursued here.

Yet another option is to work on the cognitive adequacy of logical systems themselves. This is the strategy pursued here. We tackle the computational issues of belief revision using frames and frame concepts, and thus resume a research programme originated by Minsky [87]. Even though Minsky himself was rather hostile toward logic-oriented approaches to human cognition, there is a logical and set-theoretic core recognisable in his account of frames. We explicate and further develop this core using set-theoretic predicates in the tradition of Sneed [104] and Balzer et al. [18].

A note on the infamous frame problem is in order here. In an investigation of frames, one would expect to find a thorough discussion of this problem. Fodor [57, 58] makes much out of the frame problem, but is charged with not knowing 'the frame problem from a bunch of bananas' by Hayes [72]. In fact, the account that

---

[4]Thanks to Hans Rott for this point.

[5]Focusing on worst-case scenarios means that the computational complexity of a problem is determined by the maximal number of computation steps that are needed to solve any possible instance of the problem.

Fodor [57, 58] gives of the frame problem is a reinterpretation of the original frame problem as described by McCarthy and Hayes [82]. While there is some substantial connection between the original frame problem and Fodor's reinterpretation, there is no need to discuss any variant of the frame problem in order to explain the computational challenges of dynamic, inferential reasoning. Likewise, there is no need to discuss abductive reasoning, as Fodor [58] does, for this purpose. Analysis of the computational complexity of belief revision gives us a more concise and less controversial exposition of the problem that dynamic, inferential reasoning is intractable in the setting of classical propositional logic. Minsky himself, in his seminal work on frames [87], makes no explicit reference to the frame problem.

1.3 Frame Logic

Why may frame concepts help reduce the computational complexity of belief changes? Such concepts have a richer structure than ordinary concepts. A telling example used by Minsky [87, p. 47] is that of a child's birthday party. Unlike an ordinary concept, this concept does not seem to apply well to a certain individual or tuples of individuals. We would not say that the concept in question applies to the birthday child, the union of birthday child and guests, or to the place where the party is given. What then are objects to which the concept of a birthday party is applied? Minsky says that it describes a situation that involves a number of different things: guests, games, presents, a birthday cake, a party meal, decor, etc. These things, normally, satisfy certain conditions: the guests are friends of the host, the games must be fun, the gifts must please the birthday child, etc.

From a logical point of view, a frame concept is a concept that applies to sequences of sets of objects as opposed to mere tuples of objects (which are not sets). Furthermore, frame concepts impose semantic constraints upon the (first-order) predicates of a small fragment of our language. For example, the guests of the birthday party are, normally, friends of the host. Frame concepts can therefore be used to interpret a piece of language in a small domain. This amounts to subdividing our global language into small sublanguages. These sublanguages, in turn, have different and yet interrelated interpretations in small subdomains. It is thus the notion of a frame concept by means of which we try to semantically explicate the notion of a cognitive module.

Modularising semantics in this way allows us to distinguish easily between intramodular and transmodular reasoning. The former type of reasoning is confined to a single module – which concerns the interpretation of a piece of language in a small domain – whereas the latter communicates information from one module to another. The distinction between intra- and transmodular reasoning gives rise to a proper logic of frames that emerges from our investigation. This logic is guided by the following two principles:

(1) Classical first-order logic remains valid within the application of a frame concept.
(2) Only atomic sentences and their negation can be inferred from one application of a frame concept to another.

Restricting the scope of applying classical logic is key to reducing the computational complexity of global inferential reasoning. Inspired by the methods of object-oriented programming, we furthermore allow compositions of frame concepts. Compositions, however, must be bounded in size so as to retain the tractability of frame logic. Note that object-oriented programming is a distinctive style of modular programming. It is commonly viewed as the greatest success story of Minsky's methodology of frames in [87].

For frame logic to be devised and investigated, some formal work has to be done. First, we show how the notion of a frame concept can be formalised using set-theoretic predicates. This will allow us to give an axiomatic account of reasoning with frame concepts. Then, we merge the set-theoretic account of frames with some AGM-style belief revision scheme.[6] The result is a belief revision theory with frame concepts. For this theory, we finally devise a truth maintenance system (TMS), i.e., an algorithm that determines how presently accepted truth values change upon new epistemic input. The TMS determines belief changes in a tractable manner.

The TMS is, furthermore, shown to serve as a powerful approximation of first-order reasoning, but it is not sound and complete with respect to first-order logic, even when confined to finite domains. Soundness and completeness, however, can be achieved for frame logic on condition of two further constraints: first, the conclusion is quantifier-free, while premises may well contain quantifiers. Second, the domain of any modular unit of reasoning is finite.

Frame logic is developed on the basis of a natural deduction system of classical first-order logic. We shall also speak of natural deduction frame logic to refer to this logic. A variant of natural deduction frame logic will be devised in the propositional resolution calculus. Resolution frame logic is shown to be sound and complete, without any further qualifications. Of course, we explain the basic concepts of the resolution calculus so as to make this investigation as self-contained as possible.

One word on unit resolution and unit propagation is in order in the context of frame logic. Unit resolution is a specific inference rule in the setting of the resolution calculus. At least one of the two premises of this inference rule must be a literal, i.e., an atom or the negation of an atom. At the semantic level, inferences licensed by unit resolution are described as *unit propagation*. Obviously, transmodular reasoning in frame logic amounts to the propagation of determinate literals to other applications of frame concepts. However, to the best of my knowledge, ideas about unit propagation have not yet been taken to the development of a proper logic of modular reasoning.

[6]The belief revision scheme of preferred subtheories by Brewka [24] proved well suited for this purpose.

1.4 TM Belief Revision

Natural deduction and resolution frame logic approximate classical first-order and classical propositional reasoning, respectively. Likewise, the TMS yields only an approximate determination of belief changes, approximate with respect to standard approaches in the AGM tradition. We turn, however, the approximate nature of the TMS into a virtue by devising a belief revision scheme that mirrors the working of the TMS. This belief revision scheme is tractable for natural deduction and resolution frame logic. Since the belief revision scheme is inspired by ideas about truth maintenance, we speak of *TM belief changes* in order to refer it. The TMS, thus, serves as a ladder by means of which we reach the two frame logics and a novel belief revision scheme.

Recall that some modification of an AGM-style approach to belief revision is necessary since a tractable logic alone does not suffice to resolve the computational issues of belief revision [89]. If we understand the notion of rational belief change in terms of AGM-style approaches, then $\mathbf{NP} \neq \mathbf{P}$ implies that there is simply no exact solution to the problem of tractable, rational belief revision. (We shall make this claim more precise in Sect. 1.8 below.) Approximations and computational simplifications of AGM-style approaches to belief revision are therefore of theoretical interest, at least from a cognitive point of view.

The semantics and the inference rules of frame logic, together with the TM belief revision scheme, are aimed at analysing our means of coping with the computational and cognitive challenges of belief revision. While we do not claim that the present account is literally true in all respects, we think that this account makes significant progress toward a logical analysis of quotidian human reasoning that is cognitively plausible. Frame logic and the TM belief revision scheme are intended to take the *cognitive turn in logic* one step further. Johan van Benthem [112, p. 67] came to speak of such a turn when reviewing specific trends in logic, such as belief revision theory and the related dynamic epistemic logics. Issues of computational complexity are explicitly mentioned as well [112, p. 74].

The first TMS was devised by Doyle [47]. For the expert reader it may be instructive to note, at this point, the differences between the present attempt at truth maintenance and Doyle's original TMS. First, the present system is more liberal concerning the logical form of what Doyle calls *justifications*. Any instance of an axiom can be a justification. Second, justifications may well become retracted. Third, there is an epistemic ranking of justifications.

1.5 Modularity in Cognitive Science

Now that we have outlined the key results of our investigation, let us relate these, in somewhat greater detail, to the modularity hypothesis in cognitive science. To a great extent, this hypothesis originated from Fodor's *The Modularity of Mind* [56].

There, Fodor develops a twofold thesis, which has been described as *minimal* or *peripheral modularity*: input and output systems of the mind, such as sensory and motor-control systems, work in a modular fashion. Central cognitive systems, by contrast, are non-modular. It is central cognitive systems that realise our capacities of explicit reasoning, belief formation and decision making. The notion of a module itself is characterised by the following properties and features in Fodor [56]: (i) domain specifity, (ii) information encapsulation, (iii) mandatoriness, (iv) fast output, (v) shallow, i.e., non-conceptual output, (vi) neural localisation and (vii) innateness.

From the perspective of evolutionary psychology, more radically modular proposals have been made, also comprising central cognition. As indicated above, this work aims to show that the human mind works in a massively modular fashion (see in particular Cosmides and Tooby [41]). According to this thesis, both peripheral and central cognition work with a modular architecture. Tractability is one of three arguments advanced in favour of the massive modularity hypothesis in Cosmides and Tooby [41]. It is the only argument we are concerned with here.

Even though this is only a very rough sketch of the modularity map in cognitive science and evolutionary psychology, it is precise enough to locate our results on this map. In devising a proper logic of modular reasoning, we aim to contribute to an understanding of modularity at the level of central cognition. The modular units of reasoning characterised by frame logic share at least two important properties with the Fodorian notion of a module: they are domain-specific and work with information encapsulation. This is good news for proponents of the massive modularity hypothesis since encapsulation and domain specifity are considered most central to this hypothesis [100, p. 63].

Let us further compare our logical explanation of a module with Fodor's notion. Elementary modules (which are not composed of other modules) have fast output insofar as they are associated with very simple patterns of inference. Furthermore, our specific notion of a module is perfectly consistent with having neural localisation. It is more than plausible to assume this property. As we are concerned with central cognition, our modules have non-shallow, conceptual output and input. Their working may or may not be mandatory. While Fodorian modules are not necessarily interactive, our logical modules are.

The present account of frames may well be viewed as a logical variant of the massive modularity hypothesis. This variant comes with a precise hypothesis about the working of information encapsulation:

(1) First-order and propositional reasoning within a module are encapsulated from extra-modular disjunctive information.
(2) Information in the form of disjunctions and implications is encapsulated in the sense that it is located within a module and that it cannot directly be accessed by other modules.
(3) Only literals can be communicated between modular units of reasoning.
(4) The elementary modular units of reasoning are given by applications of generalisations.

By a generalisation we simply mean a universal proposition, which may or may not be strictly believed. Strict belief of a universal proposition means that we believe all instances to be true, whereas non-strict belief amounts to believing that a great deal of the instances are true. Non-strict generalisations are needed for default reasoning. As a single application of a generalisation (be it strict or non-strict) only concerns a small tuple of objects, our modules of reasoning are highly domain-specific. Thus, we shall also be precise about the way in which logical modules are domain-specific. Frame logic itself is of course domain-general.

A literal is an atomic formula or the negation of such a formula. A literal is called a *gound literal* if and only if it has no occurrences of a variable. Throughout this investigation, the concept of a non-ground literal will not be used. For this reason, we shall often simply speak of literals to refer to ground literals. For example, we shall say that only literals can be inferred from one application of a frame concept to another, thus leaving the qualification *ground* implicit.

Let us briefly motivate composition of modular units of reasoning. Recall that any disjunction of literals can be translated into a logically equivalent implication, and vice versa. For example, $p \rightarrow q$ is logically equivalent with $\neg p \vee q$. So, it suffices to analyse disjunctions when analysing the complexity of reasoning with elementary modular units of reasoning. We can now distinguish between two ways of exploiting disjunctive information. First, by the Disjunctive Syllogism, and second by reasoning by cases. Here is an instance of the Disjunctive Syllogism:

$$\frac{p \vee q, \quad \neg q}{p}.$$

Likewise, we can infer $p \vee q$ from the set $\{p \vee q \vee r, \neg r\}$. These inferences can easily be captured by intramodular reasoning using only elementary modules (which are defined by a single disjunction or implication). Moreover, iterated applications of such direct inferences from a disjunction and a literal are captured by "communication" of literals between elementary modular units of reasoning.

More complex are inferences that involve *reasoning by cases*:

$$\frac{p \vee q, \quad p \vdash r, \quad q \vdash r}{r}.$$

$p \vdash r$ means that we can infer r from p, possibly in a direct way using Disjunctive Syllogism. Reasoning by cases cannot be captured by elementary modular units and their interaction. Hence, we need compositions of such modular units. The composition does not consist in a simple and straightforward interaction of modular units, but is more demanding insofar as there is no information encapsulation of the disjunctive information within a composition of elementary modules. Thus, full propositional reasoning can be exercised within a composite module. Notably, we must limit compositions, for otherwise intractability ensues.

Such are the basic ideas of frame logic. The just explained distinction between two types of exploiting disjunctive information is crucial to the present approach to modularity. However, the natural deduction system of frame logic is more involved and contains frame-relative inference rules for all logical symbols of first-order logic. The resolution variant of frame logic has the virtue of being simpler and easier to implement for automated reasoning.

Apparently, the patterns of information encapsulation and domain specifity as well as the interaction between modules can be described and summarised in a relatively simple way. Why is it then necessary to start with rather heavy tools, such as set-theoretic predicates of the Sneed formalism? With some qualifications, set-theoretic predicates and their axiomatic characterisation work as a ladder that can be thrown away once the inference rules of frame logic have been established. The natural deduction system of frame logic can be presented without set-theoretic predicates. Such predicates, however, are needed to define a proper model-theoretic semantics of frame logic. The soundness and completeness proofs for natural deduction frame logic are, therefore, based on set-theoretic predicates and an axiomatic characterisation of their semantics. The set-theoretic tools, moreover, help us understand in which ways frame logic is inferentially weaker than first-order logic, while being inferentially strong enough to serve as a reasonable approximation of this logic.

Our logical variant of the massive modularity hypothesis bears some resemblance with ideas about informational modules in Samuels [99]. Informational modules are domain-specific bodies of knowledge, leading to what Samuels calls a *library model of cognition*. Carruthers [35, p. 34] argues that evolution would replace innate informational modules by innate computational ones. He goes on to envision that general reasoning capacities are actually realised 'in cycles of operations of, and interactions between, existing modular systems' (p. 35). Applied to (non-innate) informational modules, this description fits very well the working of frame logic and the TM belief revision scheme.

Carruthers [35] and Samuels [99] diverge as to what extent, if at all, the human mind has something like a central processing unit that carries out the drawing of inferences. For the present account of tractable belief revision to be implemented, some central processing may be needed to coordinate the interaction of the modular units of reasoning. But both the internal inference mechanism of an elementary modular unit of reasoning and the transmodular transmission of literals among such units is so simple and straightforward that these types of inference may well be drawn without central processing. That is, as soon as a piece of disjunctive information receives a new literal, the corresponding inference is drawn locally (where this piece of information is stored) and mandatorily in circumstances in which any inference is in fact drawn. Central processing, however, may come into play when we encounter inconsistencies and when we start composing modular units of reasoning.

Unlike the informational modules in Samuels [99], the informational modules of frame logic are not innate. Even so, the capacity to form such informational modules may well be argued to be innate, in a manner comparable to how Chomsky [37] envisioned universal grammar to be innate. If one were to further argue that informational modules have their inference mechanisms hard-wired into the modules themselves, such an argument would lend support to the view that, at least, the minimal core of frame logic is innate. The minimal core is defined by frame logic without composition and confined to sentences that are literals or implications equivalent to disjunctions of literals.

Unlike Carruthers [35], we do not invoke 'quick and dirty heuristics' as a means to achieve tractability. To be more precise: if we understand the notion of a heuristics in opposition to a domain-general, possibly logic-oriented account of human cognition, then the present approach to tractability is not about heuristics at all. If we understand the notion of a heuristics in the weaker sense that using a heuristics means aiming at approximate solutions of a given problem, then our approach certainly works with heuristics.[7] Domain-specific heuristics may nonetheless form an important part of human cognition. But such heuristics are limited in explaining our capacity to reason about new domains of knowledge and enquiry.

Much more can be said about the present account of modularity in relation to research on modularity in cognitive science and evolutionary psychology. At this point, however, I would like to leave it to the reader to recognise and to pursue further connections. The present book is, of course, part of cognitive science since tractability is an issue only if we accept some form of CTM. One intent for writing the book has been to inspire further, more empirical research on modularity. For example, people working at the intersection of logic and psychology will find it easy to derive empirical predictions from the present account of modularity.

Let us briefly indicate these predictions. First, the present account of tractable belief revision predicts that human subjects have difficulties exploiting disjunctive information if this information cannot be exploited directly in the context of literals. For, composition of modular units of reasoning is computationally more demanding than iterated, straightforward inferences from disjunctions, even if the inferences to be compared have the same length. Second, the TM belief revision scheme makes distinct predictions as to how human subjects deviate from standard belief revision schemes in order to achieve tractability. Third, only very few pieces of everyday reasoning are such that reasoning by cases is unavoidable. For otherwise, this type of reasoning would not be fast.

1.6 Modularity in Knowledge Representation

Minsky's seminal work on frames contributed, to a great extent, to the development of object-oriented programming. The question, thus, arises whether frames can be used to advance logic-oriented knowledge representation in a manner that parallels the advancement of traditional, imperative programming by the concept of a frame

[7] The notion of a heuristics is not clear-cut at all. See Neth and Gigerenzer [90] for a brief account and some historical references. There is some consensus that the notion of a heuristics is in opposition to a logic-oriented account of human cognition.

(cf. [13, p. 275]). In devising a proper logic of modular reasoning, we answer this question in the affirmative. The TM belief revision scheme, furthermore, exploits the division of a corpus of information into modular units.

Of course, much work has been done on bringing modular design to bear on logic-oriented knowledge representation in past decades. To the best of my knowledge, however, no proper logic of modular reasoning has been devised so far that comes with a fully-fledged proof theory and semantics, and that exploits ideas about frames and modular design. Work on partition-based logical reasoning by Amir and McIlraith [3] is driven by similar ideas, but not taken to the development of a logic of modular reasoning with features like encapsulation and composition. Nor is this work aimed at approximating first-order reasoning by partitioning a knowledge base.[8]

The present book is written from the perspective of formal epistemology and cognitive science. It is driven by the spirit of early cognitive science, according to which an analysis of human cognition may well help us devise systems of automated cognition, and vice versa. That is, research on automated systems of knowledge representation may help us establish theoretical accounts of human cognition. The truth maintenance system is expounded in such a manner that an implementation should not be difficult for people working in the more applied areas of knowledge representation.

1.7 Extant Tractable Logics

Recall that the notion of tractability, as defined in the theory of computational complexity, explicates some notion of computational feasibility. Problems that are not tractable are considered computationally infeasible, in the sense that a computational device may take far too many computation steps to solve non-trivial instances of the problem within a reasonable time. Put simply, even our best computers may take years to solve moderately complex instances of decidable, yet intractable problems. In light of this, it is desirable to work with tractable logics in knowledge representation, at least when real-world applications are to be achieved in the long run. A logic is called tractable if an only if the problem of testing for the satisfiability of a set of formulas is tractable. If the latter condition is satisfied, we know that the problem of testing for logical entailment is tractable as well.

---

[8]The label *frame logic* has already been used for another logic, which however is designed to reason about objects in an object-oriented language (see Kifer et al. [76]). This logic is not aimed at bringing features of object-oriented design to bear on a logical account of quotidian and scientific reasoning. Hayes [71] attempts to give an account of the *logic of frames*, but the result is quite different from the logic of our investigation. Apologies about any confusion arising from these ambiguities. I could not think of a better label for the logic of modular reasoning that results from the present investigation.

Propositional Horn logic is by far the most prominent tractable logic considered in knowledge representation. The programming language Prolog is based on Horn logic. (*Prolog* is an acronym that stands for *programming in logic*.) The concepts of Horn logic, in turn, are motivated by the resolution calculus, which we shall use when devising frame logic. Without going further into any details, we can say that propositional Horn logic captures classical propositional reasoning with sentences of the following two types:

$$p_1 \wedge \ldots \wedge p_n \rightarrow q$$

$$l.$$

The first type of sentence is an implication such that all conjuncts of the antecedent and the consequent are propositional atomic formulas. The second type is simply a propositional literal l, i.e., a propositional atom or the negation thereof.

Propositional Horn logic is tractable, but a price needs to be paid for this. We cannot express disjunctive propositions whose disjuncts are not negated. For example, we cannot translate the proposition

$$p \vee q$$

into a Horn formula or a set of such formulas. Nor can we express a propositional implication with an antecedent of two negated conjuncts. That is, we cannot translate the proposition

$$\neg p \wedge \neg q \rightarrow r$$

into a Horn formula or a set of Horn formulas. The latter restriction is closely related to the former: if $\neg r$ was given or derivable from presently accepted premises, we could derive $p \vee q$ from $\neg p \wedge \neg q \rightarrow r$.

By means of these syntactic restrictions, Horn logic avoids reasoning by cases altogether. This type of reasoning cannot be expressed in Horn logic.[9] So, it seems as if we had to choose between the scylla of intractability and the charybdis of not being able to engage in reasoning by cases. But there are middle paths between these two undesirable alternatives. Frame logic is such a path. Reasoning by cases is available, but only within a modular unit of reasoning. For greater flexibility, compositions of such unit are admitted. Compositions, however, must be bounded in size. Otherwise, intractability ensues.

The *logic of limited belief for reasoning with disjunctive information* by Liu et al. [80] is another way to avoid the undesirable consequences in question. Tractability is achieved, in this logic, by delimiting the levels of iteration of reasoning by

[9]Disjunctive logical programming is based on various liberalisation of Horn logic (see Minker [86] for an overview). Without additional restrictions, however, this type of programming is not tractable.

cases. This method differs from our modular constraint on reasoning by cases, and is therefore not equivalent to it. Our hypothesis is that grouping pieces of information into modular units and composing such units are cognitive operations. This hypothesis is well motivated by Minsky's and related work on frames, and the arguments in favour of a modular architecture of human reasoning in cognitive science.

At the same time, it must be admitted that the present account of modularity does not specify under which circumstances pieces of disjunctive information are put together so that reasoning by cases can be performed. Hence, empirical research is needed to answer the latter question, and to say more about the concrete division of a given corpus of information into modular units. We consider this connection to empirical research a virtue rather than a flaw.

One word on description logics is in order. The key purpose of these logics is to define complex concepts using Boolean operations and relations, on the basis of a set of primitive concepts. For example, the concept of a professor may be defined using the following schema: a professor is a human who teaches classes and who supervises students. Description logics have become quite popular in past decades.[10] However, as regards the problem of tractability, not much progress has been made. Standardly expressive description logics are intractable to a higher degree than classical propositional logic is. For expert readers: they are **EXP**-time complex (see [16, Chap. 3]).

There are tractable description logics that form a family called *DL-Lite* [29]. Notably, these logics disallow reasoning by cases altogether, just as Horn logic does. That is, we have no means to define a concept using a disjunction. DL-Lite does not contain a logical symbol for this Boolean operation. Consequently, we cannot define, for example, the concept of a vehicle as follows: an object is vehicle if and only if it is a car, a truck, or a bus. Future research must show whether or not the present approach to tractability may be carried over to the framework of description logics.

1.8 Extant Tractable Belief Revision Schemes

A belief revision scheme is a specific method to revise a set of beliefs. Each scheme can be seen as a theory of how to rationally revise one's beliefs in light of new epistemic input. Such a theory is spelled out in terms of certain set-theoretic operations on sets of sentences or sets of possible worlds. These operations are always relative to an underlying logic, which is assumed to govern the beliefs of an epistemic agent. We shall say more about this in Chap. 3.

[10] See the *Description Logic Handbook* by Baader et al. [16] for a comprehensive account of the theoretical foundations, algorithms and applications.

The computational complexity of a belief revision scheme is determined by two factors. First, the computational complexity of the underlying logic. Second, the complexity of the set-theoretic operations in terms of which belief changes are defined. Hence, to devise an account of tractable belief revision, two things are needed. First, a tractable logic. Second, a belief revision scheme that is tractable for this logic.

Let us very briefly review the tractable accounts of belief revision to be found in the literature.[11] Suppose the epistemic state of an agent is represented by a set A of sentences (which are believed) and some epistemic ordering on A or sets of subsets of A. The concept of an epistemic ordering is motivated by the observation that some beliefs are more firmly established than others. Then, there is no belief revision scheme that is tractable for propositional logic. Not surprisingly, there is no belief revision scheme that is tractable for first-order logic either. There are only three belief revision schemes that are tractable for Horn logic: full meet revision, cut revision and linear revision (see Nebel [89, Sect. 10]). All of these belief revision schemes have highly counter-intuitive outcomes or rest on a highly counter-intuitive assumption about the epistemic ordering of beliefs. Let us briefly see in which ways these revision schemes are severely limited and counter-intuitive.

Ad full meet revision. Suppose we receive a new epistemic input α that is not consistent with our present beliefs, where α is a sentence. Then – according to the method of full meet revision – we need to retract all our present beliefs β that are not logically true. This is highly counter-intuitive for all beliefs β that have no connection whatsoever to the new information α. Suppose, for example, I believe there is butter in the fridge, but then find out that there is not. In this situation, I would like to hold on to numerous other beliefs, such as beliefs as to where I live, where I work, that the sun will rise tomorrow, etc. But the method of full revision does not allow for this type of conservativity.

Ad cut revision. This method is not quite as radical as full meet revision, but demands a large range of highly counter-intuitive retractions as well. If the new epistemic input α is not consistent with our present beliefs, then we need to retract all beliefs β that are not strictly more firmly established than the subset of present beliefs that implies $\neg\alpha$. To be more precise, we can hold on to a belief β only if it is more firmly established than any least firm member of any set of beliefs that implies $\neg\alpha$. The retraction of the other beliefs does not depend on any inferential connections between α and β. Suppose I believe that there is butter in the fridge and that my bicycle is in the shed. Let us assume, furthermore, these two beliefs have the same epistemic priority. Then, I find out that there is no butter in the fridge. In this

[11] I am drawing here mostly on Nebel [89], who has given a very comprehensive survey of complexity results in belief revision theory. While this survey dates back more than 20 years ago, no dramatic changes in belief revision theory are recognisable that have had an impact on the problem of tractable belief revision. I am not aware of a single tractable belief revision scheme in the AGM framework that goes beyond the tractable ones described in [89]. Hector Levesque indicated to me that tractability is rarely studied by people working in the area of belief revision theory.

situation, the cut revision scheme demands to retract not only the belief that there is butter in the fridge, but also the belief that my bicycle is in the shed. Likewise, for all other beliefs that are not more firmly established than the belief that there is butter in the fridge. In sum, the cut revision scheme "cuts away" too many beliefs.[12]

Ad linear revision. This revision scheme assumes, for any epistemic state, a total epistemic order on the set of beliefs. There cannot be two (logically non-equivalent) different sentences that are equally firmly believed. For example, if I believe that there is butter in the fridge and that my bicycle is in the shed, then I must believe one of these two propositions more firmly than the other. This constraint must hold for all pairs of beliefs of a given epistemic state. The assumption of a total – also called *linear* – epistemic order of beliefs is highly counter-intuitive. If I make a series of relatively direct observations when working in the lab, then the beliefs acquired thereby are often equally firm.

So far, we have assumed that an epistemic state is represented by a set of sentences and an epistemic ordering on this set or an epistemic ordering on the set of its subsets. There is an alternative representation of epistemic states that comes in the form of an ordering of possible worlds. The latter type of representation led to the development of dynamic epistemic logics.[13] Possible worlds representations of epistemic states have many benefits when it comes to defining a semantics of belief changes and corresponding logics. However, a possible worlds representation of an epistemic state is not computationally feasible. For, the number of possible worlds to be considered grows exponentially with the cardinality of the set of descriptive symbols of the language. If we use a propositional language with n different propositional variables, we have 2^n different possible worlds. In sum, possible worlds approaches to belief revision are computationally infeasible because the representation of an epistemic ordering is representationally infeasible from a computational and a cognitive point of view.

Nebel [89, Sect. 4.1] discusses several belief revision schemes that work with a possible worlds representation of epistemic states. These turn out to be tractable for Horn logic if the syntactic size of the sentence that represents the new information is bounded by a constant. However, the problem with these belief revision schemes is that they are not furnished with an epistemic ordering on the possible worlds. The corresponding representations of epistemic states do therefore not take into account that some beliefs are more firmly established than others. Such differences in firmness among our beliefs, however, are crucial for a reasonable representation of belief revision, as we shall see more clearly in Chap. 3. Assuming that all our beliefs are equally firm is highly counter-intuitive, and has highly counter-intuitive consequences. The model-based belief revision schemes that are tractable for Horn logic do therefore not provide us with a reasonable account of tractable belief revision.

[12]The term *cut revision* characterises a belief revision scheme that is based on an ensconcement ordering by Williams [117] (cf. Nebel [89, Sect. 6]).

[13]See van Ditmarsch et al. [113] for a comprehensive introduction.

As explained in Sect. 1.4, we shall work out a novel belief revision scheme that is tractable for frame logic. This belief revision scheme exploits ideas about truth maintenance, and is therefore called *TM belief revision*. It will be shown that the TM belief revision scheme has none of the counter-intuitive properties observed for the extant tractable belief revision schemes.

1.9 Overview

Part I begins with an introduction to frame concepts in Chap. 2. We shall also explain, in this chapter, the notion of a set-theoretic concept. Some references to object-oriented programming may deepen our understanding of frames and will prove useful when we come to deal with the challenges of truth maintenance at a later stage of the investigation. Chapter 3 introduces the fundamental concepts and principal approaches in belief revision theory. The overall rationale of defeasible reasoning is explained in Chap. 4. There, we shall introduce an inference system of defeasible reasoning in greater detail. Also, a brief account of the truth maintenance system by Doyle [47] is given, together with an explanation why this system has very interesting properties from a cognitive point of view.

Part II starts with a novel and axiomatic account of structuralist theory representation in terms of frame concepts. Chapter 6, then, introduces defeasible forms of reasoning into the structuralist approach, which paves the way for defining belief revisions in this approach. Such are the logical and set-theoretic foundations upon which the truth maintenance system will be devised in Chap. 7. Frame logic and the TM belief revision scheme, finally, emerge from the TMS and the semantics of structuralist belief changes in Chap. 8. In this chapter, the key tractability results are established.

As explained above, the TMS to be devised serves as a ladder by means of which we reach frame logic and the TM belief revision scheme. Unlike the ladder of Wittgenstein's *Tractatus logico-philosophicus* [119], there is no need to throw away the ladder once we have climbed it up. From a theoretical point of view, however, it is not necessary either to keep the entire ladder. The key tractability results for frame logic and the TM belief revision scheme will be established in such a manner that they do not depend on the demonstration that the TMS algorithm has a certain computational complexity. The study of truth maintenance, thus, is part of a heuristic process that leads us to frame logic and the TM belief revision scheme. I decided to make this process transparent in the present book for mainly two reasons. First, key concepts of frame logic and the TM belief revision scheme are essentially motivated by the TMS and the complexity analysis of its algorithm. Specifically, some concepts of the TM belief revision scheme are hard to understand without the context of the TMS. Second, people working in knowledge representation and knowledge engineering may well be interested in working out further developments, simplifications, and implementations of the TMS.

Part I
Foundations

Chapter 2
Frames

This chapter introduces the general notion of a frame on the basis of Marvin Minsky's seminal *A Framework for Representing Knowledge* [87]. An overview of the principles of object-oriented programming (OOP) follows upon this introduction. The chapter closes with a section on theory-elements in the structuralist representation scheme of scientific knowledge. Frame concepts figure essentially in both OOP and the structuralist approach. The present account thus emphasises substantial commonalities between Minsky-frames, classes in OOP and theory-elements in structuralism.

2.1 Minsky-Frames

2.1.1 The Abstract Theory

In his seminal paper on frames, Marvin Minsky contends that progress in Artificial Intelligence (AI) calls for a more careful consideration of how knowledge is organised in human cognition. Hence, he seeks to exploit ideas coming from both philosophy and psychology that appeared to have a closer connection to human cognition than the logic-oriented approaches to AI at the time. The very notion of a frame has its roots in the Kuhnian notion of a paradigm and the notions of a Gestalt and a schema in psychology. Using frames for the representation of human knowledge is motivated as follows [87, p. 1][1]:

> It seems to me that the ingredients of most theories both in Artificial Intelligence and in Psychology have been on the whole too minute, local and unstructured to account either

[1] All page references are based on the version available via the URL provided in the references.

H. Andreas, *Dynamic Tractable Reasoning*, Synthese Library 420,
https://doi.org/10.1007/978-3-030-36233-1_2

> practically or phenomenologically for the effectiveness of common-sense thought. The "chunks" of reasoning, language, memory and "perception" ought to be larger and more structured; their factual and procedural contents must be more intimately connected in order to explain the apparent power and speed of mental activities.

Minsky's account of frames is about the chunks of reasoning, language and memory that are presumed to make mental activities powerful and efficient. The theory thus starts with a cognitive characterisation of a frame, which is then complemented by a formal account. This account proved inspiring and fruitful in computer science. Let us begin with the cognitive characterisation. In essence, Minsky explains, a frame is a structure that we select from memory when we encounter a new situation [87, p. 2]. Such a structure is supposed to guide our detailed cognition of the encountered situation as well as further actions that may be appropriate or even required in that situation. A number of examples that Minsky adduces thus concern human perception and planning.

At a very general level, a frame is characterised as a data structure for representing a stereo-typed situation. The following explanation is instructive to see what is distinct of a frame data structure as opposed to a non-frame one [87, p. 2]:

> We can think of a frame as a network of nodes and relations. The "top levels" of a frame are fixed and represent things that are always true about the supposed situation. The lower levels have many terminals – "slots" that must be filled by specific instances or data. Each terminal can specify conditions its assignments must meet. (The assignments themselves are usually smaller "subframes".) Simple conditions are specified by markers that might require a terminal assignment to be a person, an object of sufficient value, or a pointer to a subframe of a certain type. More complex conditions can specify relations among the things assigned to several terminals.

In brief, propositions and slots make up a frame. Slots are constants of the data structure that can be interpreted only with values of a specific type. Each slot, therefore, must be associated with a specific data type. The slots of a frame may have default assignments that can be updated when the frame is applied to a particular situation. Such assignments are intended to capture generalisations that are valid only with exceptions. Hence, we can say that a frame consists of a set of slots that may have default assignments and whose actual assignments must satisfy certain propositions. With this formal characterisation the notion of a frame comes into the reach of automated reasoning.

That a frame is not only a data structure but is also composed of propositions concerning the values of the elements of such a structure becomes apparent from the claim that certain things are always true about every situation that the frame is supposed to represent. This statement is left without an explanation, but it has an interesting reading in our context: in addition to simple conditions that the assignments of values to slots are required to meet, there may also be complex conditions that relate slots to one another. "More complex conditions can specify relations among the things assigned to several terminals" [87, p. 2]. If one were to describe formally such complex and simple conditions, one would represent them as propositions that have occurrences of constants being labels of slots. Simple conditions have only one such occurrence, whereas complex ones must have at least two.

Once the notion of a proposition about slot values is in play, it makes sense to distinguish between interpretations of slots satisfying such a proposition and those that do not satisfy it. Only those interpretations that satisfy the propositions that are essentially associated with a frame are instances of that frame. That is, we suggest, a sensible reading of the claim that certain things are always true about every situation that the frame is supposed to represent. Note that Minsky describes the event that a 'frame cannot be made to fit reality' as a case where no assignment of values to slots can be found that satisfies the frame's assignment conditions [87, p. 3]. This further justifies the view that a frame consists of slots and propositions concerning the admissible values of these slots.

Another important element of the formal account is that frames are related to one another, thus making up a frame system. Relations among frames are represented also by means of slots insofar as the data-type of a slot within a frame can be another frame. Relations among frames may fulfil several types of functions: they can represent transformations between perspectives in the case of visual perception, causal relations, means-ends relations as well as explanations and justifications. Moreover, sequences of frames may guide corresponding sequences of enquiry when, for filling the slots of subframes, further information is required. Frame systems may form a *similarity network* such that in case of a mismatch between a proposed frame and reality an algorithm gets started that searches for a more fitting frame within the network. We shall use relations among frames to represent justifications and explanations within a formal theory of belief changes.

From a philosophical perspective, it is worth noting that Minsky sees his abstract theory of frames as standing in the tradition of *paradigms* in Kuhn [77] and *schemas* in Bartlett [19]. The works of the *Gestalt theorists* are cited frequently, moreover. One, therefore, can view the notion of a frame as an attempt at a formal elaboration of the interrelated ideas of a paradigm, a schema and a Gestalt.

2.1.2 Examples

The abstract theory of frames is illustrated with detailed examples from different fields, such as visual perception of spatiotemporal objects, grasping the grammatical structure of a sentence and understanding causal relationships in a story. Let us have a look at one of these examples to get a more vivid conception of frames in the sense of Minsky. The event of attending a children's birthday party may be captured by the following frame with default assignments [87, p. 47]:

Dress – Sunday best.
Present – Must please host. Must be bought and gift-wrapped.
Games – Hide and seek. Pin tail on donkey.
Decor – Balloons. Favours. Crepe-paper.
Party-meal – Cake. Ice-cream. Soda. Hot dogs.
Cake – Candles. Blow-out. Wish. Sing birthday song.
Ice-cream – Standard three-flavour.

Arguably, in this example, default assignments are somewhat mixed with simple conditions that the potential fillers of slots are required to satisfy. Whereas the slot for games has been assigned to a default value, viz., a set whose members are particular games, the slot for a present is rather associated with a simple condition. The example of a frame for a birthday party is embedded in a larger story about preparations for such a party. In the course of this story, relations among frames are intended to guide a train of actions.

A slight variation of the birthday party frame (not given by Minsky himself) helps us exemplify the distinction between simple and complex conditions for slot assignments:

Host – Has birthday. Receives presents.
Guests – Are friends of the host. Give presents to the host.
Presents – Must please the host.
Games – Host and guests participate. Being fun.

The division of assignment conditions of slots into simple and complex ones should be obvious in this example. Whereas *being fun* is a simple condition for games, *host and guests participate* is a complex one for the games slots. It is only complex conditions that relate the filler of a slot to another. Note that in this example slots may have both individuals and sets of individuals as assignments.

Even though this example by Minsky is rather simple, it may be impressive enough to show that frames have a richer structure than ordinary first-order concepts. Frames apply to tuples of objects, where some of these objects may be sets of objects. This is not admissible for first-order concepts, provided the first-order language is not enriched by an axiomatic set theory. Another difference is that frames may serve as values of a slot within a frame.

Can we think of frames as a peculiar type of concept having richer structure than first-order ones? Arguably, we can. This, at least, is the viewpoint from which the notion of a frame has been introduced here. The instances of a frame, which are given by tuples of objects, are related to the frame itself in precisely the same way as object and concept are related to one another in predicate logic. This understanding of frames is supported by passages where Minsky himself speaks of *frame concepts* [87, p. 52]. It seems promising, therefore, to think of frames simply as a particular species of concepts in higher-order logic. This perspective will be adopted in Sect. 2.3 of this chapter.

2.1.3 Criticism of the Logistic Approach

As Kuhn's analysis of paradigms in science breaks with the standard account of scientific theories, so is Minsky's proposed framework for representing knowledge intentionally moving away from the logic-oriented approach to Artificial Intelligence. The primary objection to this approach is that the items of reasoning, language and memory are too minute to account for the efficiency of human reasoning. In the last section of his seminal paper [87, pp. 100–106], Minsky is more precise about the limitations of logical languages. It is worth investigating which of these limitations have been overcome in novel logics developed since then and which remain to impede the use of logical systems for representing knowledge. Logical systems are found deficient by Minsky in the following respects.

First, there is the issue of relevance. Commonsensical and scientific reasoning is *selective* in that only certain axioms and inference rules are applied to only a subset of the whole set of items of presumed knowledge. In other words, only a very minor fraction of the whole range of proper inferential transformations are considered relevant by human agents. In a knowledge base that consists of atomic propositions and general axioms, however, no information of relevance is encoded. What is called for, Minsky argues, is an *inter-propositional* schema of representing knowledge that joins items of knowledge with information about how to use these items.

Second, inferential patterns used in commonsensical and scientific reasoning are rarely truth-preserving in the strict sense of classical deductive logic. Rather, we rely on some notion of truth-preservation under normal conditions, thus intentionally ignoring some unusual and yet conceivable scenarios.

Third, there is the well-known problem of combinatorial explosion of theorem proving algorithms. There is no complete and correct algorithm known to us that determines logical implication in propositional logic without having exponential computational complexity. First-order logic is not decidable, i.e., there is no algorithm that determines whether a given formula is a theorem of first-order logic. Textbook examples of logic-oriented AI techniques may work well, Minsky argues, but as soon as one extends the knowledge base, one is facing serious computational problems.

Fourth, logical systems need to be consistent and are intended to be complete in the sense that any true proposition can be derived from the knowledge base. Neither the former nor the latter is considered a desideratum by Minsky. The argument against consistency is plain and convincing: no human mind is ever completely consistent [87, p. 104]. Rather, we have adopted certain strategies to handle contradictions. For example, if a proposition and its negation can be derived from a set of accepted premises, then this proposition and its negation should simply not be used for further derivations. It is the *selective* use of inferential patterns that prevents human reasoning from breaking down in the face of contradictions. In classical and many non-classical logical systems, by contrast, every proposition becomes derivable once the premises contain a contradiction. Minsky's argument against completeness is less clear, to my mind. At the core of this argument seems to be the claim that logical systems do not have to be complete in applications: if certain consequences are of no interest in the application, then there is no need to capture these consequences by the theorem prover in use.

Which of these problems have found a satisfying solution through the advancement of new logics in past decades? There does not seem to be a logical system that allows for inter-propositional knowledge representation. One may even argue that there is no way to make a logical system inter-propositional. If we were to attempt this, we would move away from the declarative style to the procedural style of programming, thus giving up the idea of representing knowledge with a logical system. This line of thought will be questioned when we come to deal with the structuralist approach in Sect. 2.3.

What about the notion of truth-preservation under normal conditions in modern logics? Clearly, we have now a representation of this in the form of a variety of nonmonotonic logics. These will play an important role in the development of our structuralist approach to belief revision in Chap. 6.

Combinatorial explosion. This problem has remained unresolved for nonmonotonic logics and belief revision. Also, no efficient algorithm is known to determine logical implication in propositional logic. It is only certain fragments of propositional and first-order logic for which efficient algorithms could be devised. A case in point is reasoning with Horn formulas, i.e., formulas that are disjunctions of literals, where at most one of these literals is unnegated. However, belief revision and nonmonotonic reasoning with Horn formulas is computationally feasible only on very restrictive assumptions. We shall be more explicit about these computational aspects in Chap. 7.

Consistency. Well, we have now paraconsistent logics and had such even at the time when Minsky [87] appeared.[2] It is open to doubt, however, whether the divergence of paraconsistent logics from the canons of classical reasoning is tolerable or not. Certain inferential transformations which cease to be valid in paraconsistent logics may be needed in non-contradictory contexts. The strategy of dealing with inconsistencies that Minsky envisioned was a different one. Whereas paraconsistent logics (in the narrow sense of systems with a formal proof theory and semantics) modify the inferential canon of classical logic, Minsky suggested that the stock of logical axioms and inference rules be used selectively: if certain applications lead to inconsistencies and paradoxes, withdraw them. Where no inconsistency is threatening, the stock of logical axioms and inference rules may be used unmodified. This is precisely the strategy pursued in the construction of *preferred subtheories*, which underlies the approach to defeasible reasoning by Brewka [24]. This construction will be explained in greater detail in Chap. 4.

In sum, the problems of relevance and of combinatorial explosion remain most pressing in logic-oriented approaches to commonsensical and scientific reasoning. Both of these two problems are addressed via the tractable belief revision scheme and frame logic, emerging from the present investigation.

2.2 Classes in Object-Oriented Programming

2.2.1 The Notion of a Class

Upon the appearance of Minsky's seminal paper and related work by others, a number of frame-like and frame-based representations of knowledge have been developed. Among them, there are frame systems in the narrow sense

---

[2] For an overview see Priest [93].

whose reasoning mechanisms are based on inheritance, procedures that get started when slots are filled and procedures for requesting values of slots [22, pp. 136–151]. Moreover, description logics [17] are considered frame-like representations.

Arguably, the most tremendous success of the frame idea has bee achieved with object-oriented programming languages, such as SmallTalk, C++ and Java. In the context of our investigation, a brief review of the basic principles of OOP helps us understand the idea of an inter-propositional representation of data and knowledge, that is, a schema in which propositions are more closely associated with information about how to use them. This association is brought about by a division of what logicians would call the descriptive vocabulary of the program unit into subvocabularies. Such a subdivision will prove a core feature of the structuralist approach and, therefore, deserves our attention.[3]

According to a common textbook explanation, the object-oriented style of programming consists in overturning the separation of procedures and data structures. In the non-object-oriented style, (global) data could be accessed and manipulated at any place of the program. Introducing variables for data of specific types was one thing and writing code operating on such data another. As a consequence, it was always time-consuming to investigate which data get manipulated at which parts of the program in what specific ways. Programming code was difficult to read and expensive to maintain; reusing code was risky since there were no explicit means to display the relevance of data structures, i.e., their intended use. Once data structures and procedures operating thereupon are more closely associated, as required by OOP, the situation becomes different.

The first step on the way to appreciate the object-oriented style of programming may be to understand the grouping of data into *objects*. The thought is that there are different types of objects and each type of object comes with a specific set of data being processed. Unlike individuals in predicate logic, objects in OOP have an inner (data) structure. A simple example may illustrate the point. In a library resource management system, the concept of a circulating item may be a sensible object type. With this object type, the following types of information may be associated: (i) days allowed for use; (ii) the date when the object is checked out; (iii) the date when the object is due; (iv) the ID of the borrower; (v) the renewal number; and (vi) the copy number. In the programming language C (which became object-oriented with its extension C++), this association of data would be introduced by means of an entity that is called a *struct*[4]:

[3] The following account of OOP is based on [26, 51, 52, 84], which are also recommended for further reference.

[4] The library resource management example is adopted, with some modifications, from [84, p. 217–219].

```
struct CirculatingItem
  {
   int      daysallowed;
   Date     dateOut, dateDue;
   int      borrowerID;
   int      renewalNumber;
   int      copyNumber;
  };.
```

The introduction of structs in C, or records in Pascal, allows one to group facts in terms of the object they pertain to. More precisely, a struct has several *instance variables* that are waiting to be assigned values, which then represent presumed facts concerning the respective object. In the present example, *daysallowed*, *dateOut*, *dateDue*, *borrowerID*, *renewalNumber*, *copyNumber* are instance variables. Such variables usually have a particular data type.

Apparently, the instance variables of a struct are nothing but slots in the sense of Minsky's frame conception. The struct itself figures as a frame of these slots. Notably, the procedure of introducing structs can be iterated, i.e., the data type of an instance variable of a struct can be another struct.

The next step on the way to object-oriented programming is to group procedures around objects such that objects become associated with the appropriate operations on their data. The unit that associates structs, or records, with procedures is called a *class* in C++ and Java. A class is nothing but a template for creating objects. The "creation" of an object is the event of allocating memory for storing values of its instance variables. Objects are created in a similar vein as the elements of a struct are defined. Here is an example of creating instances of the object type CirculatingItem:

```
vector<CirculatingItem>   ci.
```

This creates a vector of CirculatingItem.[5]

To show how the association of procedures with structs works, let us continue with our library example. There are two obvious operations that one might want to perform on a circulating item, renewal and handing it back. In C++, the class definition that introduces these operations may look as follows:

[5]Loosely speaking, a vector is a one-dimensional array of dynamic length.

```
class CirculatingItem
  {
   protected:
     Date      dateOut, dateDue;
     int       borrowerID;
     int       renewalNumber;
     int       copyNumber;
     bool      available;

   public:
     void  borrowIt (int ID, Date today)
       {
        dateOut = today;
        dateDue = today.mday + 28;
        dateOut.mkDate;
        borrowerID = ID;
        renewalnumber=1;
        available = falsc;
       }
     bool  renewal (Date today)
       {
        If (renewalNumber < 3)
          {
           dateDue.mday=today.mday+28;
           DateOut.makeDate;
           renewalNumber++;
           return true;
          }
        else return false;
       }
      void  handingBack( )
       {
         available=true;
         borrowerID=0;
       }
   }.
```

The behaviour of the functions *borrowingIt*, *renewal* and *handingBack* is as follows. Calling *borrowingIt* takes the ID of the borrower as an argument and sets the instance variables of the object to appropriate values. If *renewal* is called and the number of renewals is less than 3, the item gets renewed, which means that the value of *dateDue* is set to the current date plus 28 days. It is assumed that *Date* is a class with instance variables for day, month and year, and that *mkDate* is an associated function that adjusts the values of day, month and year in case any of these values are beyond the admissible range.[6] Calling the function *handingBack* has the effect that the instance variable *available* is set to true.

[6]This function is analogous to *mktime* of the *time* library in C++.

A function may have both a return value and a tuple of arguments. *BorrowIt* is a function that takes a value of type *integer* and another value of type *Date* as arguments but has no return value. *handingBack*, by contrast, has a return value. This value is of type *bool* and indicates whether or not the request to renew was possible or not. Return values must be declared in front of the function identifier. The return type *void* indicates that no value is returned. Obviously, the concept of a function in C++ and the concept of a function in mathematics and logic differ slightly from one another, but this should not cause much confusion.

In C++, the instance variables of a class are called *member fields* and the procedures of a class are called *member functions* of that class. This terminology will be adopted henceforth. Member fields are accessed with commands of the form $object\_identifier.variable\_identifier$, and member functions are called by a command of the form $object\_identifier.function\_identifier$ $(argument_1, \ldots, argument_n)$. The meaning of the keywords *protected* and *public* will be explained below.

A note on the different use of the term *variable* in logic and in conventional programming is in order. In predicate logic, variables are exclusively used in the context of quantifiers. We need variables to express that all objects, some object, exactly one object, etc. satisfy a certain property. This is not so in conventional programming. There, a variable is a particular type of symbol that receives an interpretation in the form of an assignment of values during program execution. In this respect, variables in the sense of conventional programming languages resemble the non-logical constants of a logical language rather than the variables of such a language. The assignment of values to a variable in conventional programming may undergo further changes during program execution, however, which gives rise to speaking of variables. Following Carnap, we shall also call the non-logical symbols of a logical language *descriptive*, for stylistic variance.

2.2.2 Composition and Inheritance

So far, objects have been presented as isolated, monadic items. According to the preceding explanations, a program seems to be composed of class definitions, commands to create objects which instantiate those definitions and commands that call methods of particular objects. An important feature of OOP, however, is to allow for "communication" between objects. This feature is enabled through the *composition* of classes within another class. Composition has already been used tacitly in the above example. The instance variables *dateOut* and *dateDue* of the class *CirculatingItem* are objects of the class *Date*, which comes with associated functions such as *makeDate*.

In more general terms, composition means that classes may be used for the type definition of instance variables within another class. For composition to work, it is necessary that the types of instance variables are not confined to built-in types, i.e., types that are built into the language from the start. *Integer*, *float* and *string* are common built-in types. If composition is used, the functionality of the component class, i.e., its data structure and methods, becomes available to the functions of the encompassing class. Composition can be seen as an n-ary relation among classes: if a class a is a composite class, it has component classes $b_1, \ldots, b_{n-1}$. These may, in turn, be composed of further classes. Composition was envisioned by Minsky when he introduced the notion of a subframe [87, p. 2]. As a class can have instance variables whose type is another class, so can the type of a slot within a frame be another frame. Such subordinated frames are simply called *subframes*.

The statement that there is communication between objects is to be understood as follows. Once a class of type b is used to define the type of an instance variables of a class a, the (public) member functions of b can be called within the member functions of a. The request to call a method of a b object made within the methods of an a object is called, sending a message to a b object. The distinction between *private* and *public* fields, as well as the parallel distinction between private and public member functions, becomes relevant here. Only public fields and public functions of a class b may be accessed or called from within the member functions of a class other than b. Private member fields and member functions of an object, by contrast, can only be accessed and used from within functions of that object.

Besides composition, there is another equally important means to associate classes, viz., *inheritance*. The idea is that the class definition of a *derived class* inherits all of the member fields and member functions of its *base class* and does, in addition, extend those members by further fields and functions. Also, the modification of member functions of the base class is permitted through redefining such functions in the derived class. Notably, the introduction of derived classes can be iterated. In C++, even multiple inheritance is admissible, i.e., a derived class may inherit functions and fields from more than one base class. Here is an example of inheritance in the form of a suitable base class for the class *CirculatingItem* in the library resource management system:

```
class OwnedItem
  {
   protected:
    char     title[81];
    float    cost;
    int      condition;
    char     ISBN[15];
    char     vendor[51];
    int      numOfCopies;
   }.
```

The class *OwnedItem* has the following instance variables: *title*, *cost*, *ISBN*, *vendor*, *numOfCopies* and *available*. Again, the meaning of these variables should be self-explanatory.

It is time to explain the keyword *protected*, which has already been used in the examples. Protected access means that only derived classes have the permission to access the member fields and to call member functions of the respective class. This kind of access is intermediate between *private* and *public* access.

To indicate that *CirculatingItem* is a derived class of *OwnedItem*, only the first line of its definition needs to be modified as follows:

```
class CirculatingItem: OwnedItem.
```

In most libraries it is not the case that every item is circulating. Encyclopedia, textbooks and editions of classical authors may be available only in a non-lending collection. Both the class *CirculatingItem* and a class for non-circulating items can inherit the data structure from the class *OwnedItem*. This is the motivation for having a separate class definition for circulating items, which is derived from the base class for owned items.

To take this example further, suppose the circulating items are books, collections, or DVD's. All these types of items may be subject to the same borrowing regulations. Among the items of the non-lending collection, there may only be books and periodicals. In this case, it is sensible to have a more general class for circulating items that groups together member fields and member functions needed for the process of borrowing. In addition to this more general class, we have then specific derived classes for books, collections and DVD's, respectively. Figure 2.1 depicts the inheritance relationships among the classes considered with their corresponding data structures.

It is assumed that movies have only one director, which requires to view Joel and Ethan Coen as one person. Books and proceedings, by contrast, may well have two or three authors or editors, respectively. Hence, a two dimensional array of *char* is defined as an instance variable for their names.

To finally give a simple example of composition, we may create a class *name* with member fields for last name, first name and middle initial:

```
class Name
  {
   private:
    char      nameLast[15];
    char      nameFirst[15];
    char      nameMidInit;
  }.
```

When used as a component class in the classes book, collection and DVD, this class allows for a more accurate and convenient representation of the names of authors, editors and directors.

It goes without saying that we touched upon only the most elementary features of object-oriented programming. An extended discussion would have to include further topics, such as constructors, defaults, polymorphism, method overloading and templates. These topics, however, are not necessary for an understanding of the use that is made of the frame idea in the present investigation.

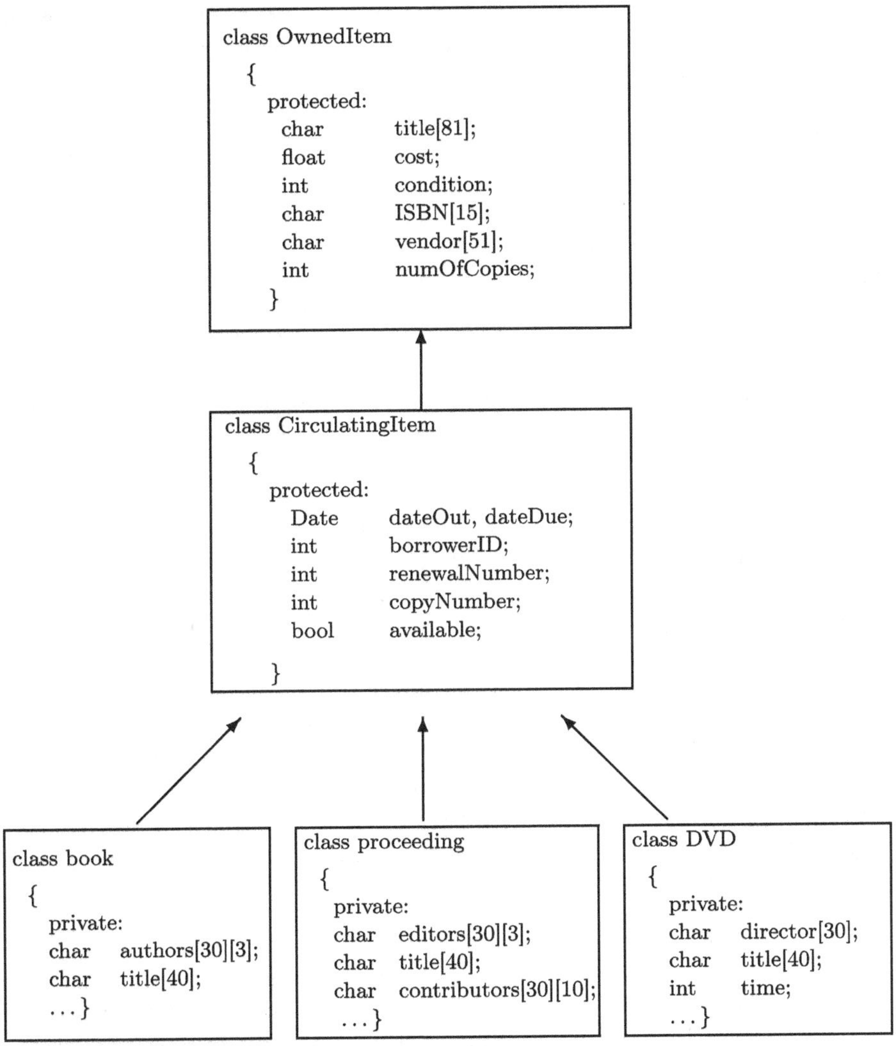

Fig. 2.1 Inheritance relationships exemplified

2.2.3 Philosophical Logic and OOP

Why deal with object-oriented programming in the context of philosophical logic? We have seen that OOP can be traced back, in part, to philosophical ideas about the patterns of scientific reasoning and to psychological theories about our means of object recognition. This perspective has been adopted here by claiming and arguing that Minsky's notion of a frame is a formal elaboration of the interrelated ideas of a paradigm, a Gestalt and a schema. There is thus some philosophical contribution recognisable in the work of Minsky, which in turn inspired the development of OOP.

Let us now take a closer look at potential connections between philosophical logic and OOP. Are there any features of this style of programming that may be fruitfully adopted for the construction of formal logical systems? When comparing logical with imperative programming languages, it is a natural question to ask whether and, if so, to what extent the distinction between logical and descriptive vocabulary applies to programming languages at all. Arguably, there is in fact a related distinction in such languages. On the one hand, we have keywords whose semantics is defined by the programming language itself. Think of the keywords *class*, *public* and *integer*. On the other hand, we have identifiers for variables, classes and functions that are introduced and interpreted during program execution. Keywords, such as *class* and *public*, are therefore comparable to the logical vocabulary of a logical language because their semantics is as rigidly fixed as the semantics of the logical constants is in a logical language. Identifiers for instance variables, functions, objects and classes, by contrast, resemble the descriptive vocabulary of a logical language.

Both variables in a non-OOP imperative programming language and instance variables of a class in an OOP language are introduced for the purpose of assigning values to them. Such an assignment of values can be thought of as an ongoing and dynamic interpretation of the descriptive vocabulary of a logical language. For, an interpretation of a logical language is likewise a mere assignment of values (of the appropriate type) to the language's descriptive constants. Admittedly, there is also a number of differences between these two kinds of assignments. The most important one is grounded in the fact that the interpretation of descriptive symbols of a logical language is static since these symbols are interpreted all at once. (Even in dynamic epistemic logic and dynamic doxastic logic, the interpretation associated with possible worlds is not envisioned to change.) This contrasts with the assignment of values to variables in programming languages, which is dynamic because it is brought about during program execution and may change during that execution.

Needless to say, the standard semantics of logical constants differs in kind from the type of semantics commonly advanced for the built-in commands of conventional programming languages. The latter is given in terms of transition systems that describe changes of the computation device that correspond to the execution of commands.[7] Such transition systems have little in common with the standard, truth-conditional semantics of logical constants.

Let us focus, however, on semantic properties where conventional programming and logical languages converge, viz., the assignment of values to variables and the interpretation of the descriptive vocabulary of logical languages. How does the behaviour of variables in OOP differ from those in non-OOP languages? Now, any variable in OOP must be an instance variable of a class, which has the consequence that it can only be accessed by methods of that class and classes inheriting from that class. In non-object-oriented languages, by contrast, variables are usually global ones, set aside local and auxiliary variables defined in Pascal's procedures and C's

[7]For details see [53].

functions. Being global of a variable means that it can be accessed and modified from within any part of the program. This difference in the behaviour of variables in OOP and non-object-oriented languages is of crucial interest: the feature that comes into play with the object-oriented style and that we seek to exploit here is that the descriptive part of the language becomes divided into sublanguages. A class definition introduces a descriptive vocabulary that forms only a fragment of the whole, global language. With the creation of an object, the descriptive vocabulary is interpreted. That is, the member fields, which consist of instance variables, are values assigned.

It is worth noting, furthermore, that a sublanguage of a class may receive more than one interpretation since a class is only a template for creating objects. Each object comes with its own interpretation of the instance variables of the respective class, i.e., the class of which the object is an instance of. Finally, by means of composition and inheritance, the sublanguages corresponding to classes can be combined to form more complex and thus more expressive sublanguages. The division of the global language into sublanguages, or alternatively, the construction of an encompassing global language out of sublanguages, has the important consequence that any operation on data must be relativised to an appropriately chosen fragment of the global language.

Let us transfer these ideas to logical languages. Propositions are then to be formulated in sublanguages of varying complexity, and the inferential representation of reasoning is to be relativised to sublanguages. Moreover, it is conceivable that one and the same sublanguage is interpreted in different domains, just as a class is instantiated a number of times. Yet, there may be means to reason about these different interpretations in a language that contains the descriptive vocabulary of the respective class. There may also be means of higher-order reasoning that refer to the interpretations of different sublanguages insofar as there is communication among different classes in OOP.

Again, we must ask ourselves what the philosophical interest is of decomposing the language into sublanguages and to relativise inferences to sublanguages. There are at least two motivations for this. First, since the frame idea has some roots in Kuhnian philosophy of science, there is the prospect of formally accounting for certain features of scientific reasoning that were emphasised by Kuhn and disregarded by Carnap. Second, arguably, once we embrace the *cognitive turn* in logic, we must wonder if there are specific logical means in terms of which human minds deal with the challenges of computational complexity in quotidian and scientific reasoning. The division of the global language into sublanguages and an object-oriented division of their interpretations will serve as point of departure for devising frame logic in the present investigation. This logic will be shown to be both reasonably expressive and tractable.

Let us give a preliminary justification of why subdividing a given language may lead to a tractable system of reasoning. The computational complexity of automated reasoning strongly depends on the cardinality of the set of the descriptive vocabulary, on the size of the knowledge base and on the cardinality of the domain of interpretation. Hence, if it turns out that, for some domains of interest, the language is decomposable into sublanguages in the style of class definitions and that the application of inferential patterns are performable within corresponding sublanguages, this would result in a significant reduction of the computational complexity of the reasoning procedures. Of particular interest in the context of belief revision is the test of whether or not a piece of presumed knowledge can consistently be added to the knowledge base. If the addition can be made in a consistent fashion, we are interested in what further consequences are derivable from the thus expanded knowledge base. In the case of an inconsistency, we need to survey the revisions through which consistency may be reattained.

In sum, object-oriented logical languages would be languages that are composed of sublanguages. These sublanguages can be interpreted independently of one another. They are combinable to form more complex languages. Expressing propositions in sublanguages and performing reasoning within such sublanguages may be key to understanding how human minds are capable of keeping the computational complexity of reasoning procedures at a moderate level. Of course, it does not suffice to simply decompose the language into sublanguages and to sort propositions accordingly. Rather, it must be shown that there are inferential patterns performable within the sublanguage chosen, other inferential patterns and propositions that concern the different interpretations of one and the same sublanguage and, finally, inferential patterns that need to be performed within a language being composed of appropriately chosen sublanguages. Only then can the representation of reasoning procedures be achieved in an object-oriented style. In the next section, it will be shown how the division of a global language into sublanguages and an object-oriented multitude of interpretations of such sublanguages is realised in the structuralist approach to scientific theories.

2.3 Theory-Elements in Structuralism

2.3.1 Set-Theoretic Structures and Predicates

The core idea of structuralism is to represent empirical systems by means of sequences of sets and to model the application of scientific theories by means of set-theoretic predicates. The use of set-theoretic predicates, therefore, distinguishes the structuralist approach from other formal accounts in the philosophy of science. In this chapter, we shall give an overview of the elements of structuralist theory representation. It will be shown that the basic concepts of structuralism qualify as frame concepts in the sense of Minsky. Let us begin with a mathematical example of a set-theoretic predicate [109, p. 250]:

Definition 2.1 (Preorder) $\mathcal{A}$ is a preorder if and only if there is a set A and a binary relation R such that $\mathcal{A} = \langle A, R \rangle$ and

(1) $R \subseteq A \times A$
(2) $\forall x \forall y \forall z (R(x, y) \wedge R(y, z) \rightarrow R(x, z))$
(3) $\forall x R(x, x)$.

The concept of a preorder applies to ordered pairs of sets that consist of a domain A of interpretation and a relation R interpreted in A. This style of introducing set-theoretic predicates is adopted in structuralism for the representation of scientific knowledge. A set-theoretic predicate is simply a predicate that applies to sequences of sets. Such a sequence is assumed to consist of a subsequence of base sets and another subsequence of relations:

$$\langle D_1 \dots D_k, R_1 \dots R_n \rangle$$

where $D_1 \dots D_k$ are base sets and $R_1 \dots R_n$ relations. Following the terminology of Bourbaki [21] and common usage in model theory, sequences of this type are called *set-theoretic structures*. As structures in model theory of formal logic specify a domain of interpretation and an interpretation of the non-logical symbols, so are the base sets $D_1 \dots D_k$ to be understood as domains of interpretation and the relations $R_1 \dots R_n$ as interpretations of corresponding relation symbols. Hence, we can say that a structure of type $\langle D_1 \dots D_k, R_1 \dots R_n \rangle$ specifies the interpretation of the relation symbols $\ulcorner R_1 \urcorner \dots \ulcorner R_n \urcorner$. Unlike structures in model theory, the interpretation of individual constants remains implicit in the case of structures of the type $\langle D_1 \dots D_k, R_1 \dots R_n \rangle$.

We use the symbols $R_1, \dots, R_n$ of a structure $\langle D_1 \dots D_k, R_1 \dots R_n \rangle$ either autonymously or to refer to their proper intended interpretation. It will be clear from the context, in most cases, which type of use is intended. If disambiguation is needed, Quine's brackets $\ulcorner \dots \urcorner$ will be used to refer to symbols. When speaking of a set-theoretic structure, we shall always mean a sequence of sets consisting of a non-empty subsequence of base sets $\langle D_1 \dots D_k \rangle$ and a non-empty subsequence of relations $\langle R_1 \dots R_n \rangle$ over these base sets.

Set-theoretic predicates specify set-theoretic structures by means of imposing constraints upon the sets that are admitted as values in a certain species of such structures, as has been shown with the above example of a preorder. One distinguishes three types of such constraints, *typifications*, *characterisations* and *laws*. A typification is a statement of the form $R_i \in \sigma(D_1, \dots, D_k)$, where $\sigma(D_1, \dots, D_k)$ stands for a sequence of concatenated operations on the base sets $D_1, \dots, D_k$. The types of operations are selection, Cartesian product and power set. A typification thus indicates that a relation R_i is of a determined set-theoretic type over the relations $D_1, \dots, D_k$.

Characterisations and laws are expressed by formulas applying to set-theoretic structures. Such formulas are built up from the symbols $D_1 \dots D_k, R_1 \dots R_n$ as well as logical and set-theoretic symbols. A *characterisation* is a formula that contains (besides set-theoretic and logical symbols) exactly one relation symbol R_i, $1 \leq i \leq n$. A *law*, by contrast, is a formula that establishes some universal, non-trivial connection between at least two relations of $R_1 \dots R_n$. The law must have occurrences, therefore, of at least two symbols for relations. To indicate that the law is non-trivial, we also speak of the *substantial law* of the *theory-element* **T**. A substantial law in this sense may consist of more than one formal axiom. Theory-elements are the units by means of which a scientific theory is composed in the structuralist approach.

2.3.2 Potential and Actual Models

It has been observed, among others, by Weyl [116] and Carnap [33] that theory formation goes hand in hand with concept formation. That is, the exposition of a scientific theory comes with the introduction of concepts that are specific to that theory. Such concepts are called **T**-*theoretical* in structuralism, where **T** stands for the theory or theory-element through which the concepts are introduced. Paradigmatic examples of **T**-theoretical concepts are mass and force in classical particle mechanics. Those concepts, by contrast, which are used to describe the empirical systems to which **T** is applied are called **T**-*non-theoretical*. The notion of theoreticity will concern us in greater detail in Chap. 5.

The distinction between **T**-theoretical and **T**-non-theoretical concepts gives rise to the following distinction between two kinds of set-theoretic entities:

$$\langle D_1 \dots D_k, N_1 \dots N_p \rangle \tag{2.1}$$

$$\langle D_1 \dots D_k, N_1 \dots N_p, T_1 \dots T_q \rangle . \tag{2.2}$$

Structures of type (2.1) are used to represent empirical systems to which **T** is applied, whereas structures of type (2.2) represent **T**-*theoretical extensions* of structures of type (2.1). The extension simply consists in a valuation of the **T**-theoretical relation symbols. The symbols $N_1 \dots N_p$ thus designate **T**-non-theoretical relations, whereas $T_1 \dots T_q$ designate **T**-theoretical ones. The symbols $D_1 \dots D_k$ designate sets of empirical objects that make up the empirical system to which the theory is applied.

Why are there different sets $D_1 \dots D_k$ of empirical objects in place of just one domain D of interpretation? This allows for a more fine-grained characterisation of the empirical systems to which a theory-element is applied, as will be shown with a number of examples throughout this investigation.

If the theory involves some mathematical apparatus, symbols for sets of mathematical objects need to be introduced as well. This results in structures of the following types:

$$\langle D_1 \dots D_k, A_1, \dots, A_m, N_1 \dots N_p \rangle \tag{2.3}$$

$$\langle D_1 \dots D_k, A_1, \dots, A_m, N_1 \dots N_p, T_1 \dots T_q \rangle . \tag{2.4}$$

$A_1, \ldots, A_m$ are sets of mathematical objects. Some or all of the **T**-non-theoretical and **T**-theoretical relations may be functions, i.e., binary many-to-one relations. In the theories of physics, most quantities are introduced as functions that take empirical objects as arguments and that have mathematical objects as values. Think of the concepts of temperature, pressure, mass, force, electromagnetic field, etc. By allowing the subsequence of sets of mathematical objects to be empty, structures of type (2.1) and (2.2) are, respectively, special cases of structures of type (2.3) and (2.4).

It is time to deal with some examples. A simple law commonly found in textbooks of classical mechanics is the lever principle. The theory-element **LP** covers the case where the weights on either side of the lever are in equilibrium [104, p. 11]:

Definition 2.2 (Model of LP) x is a model of the lever principle ($x \in \mathbf{M}(\mathbf{LP})$) if and only if there exist D, n, t such that

(1) $x = \langle D, \mathbb{R}, n, t \rangle$
(2) D is a finite, non-empty set
(3) $n : D \to \mathbb{R}$
(4) $t : D \to \mathbb{R}$
(5) $\sum_{y \in D} n(y) \cdot t(y) = 0.$

n has the intended meaning of the distance function from the centre of rotation of the lever and t the intended meaning of the mass function. Among the elements of this definition, we can distinguish between laws and characterisations. (2)–(4) are characterisations, whereas (5) clearly qualifies as a law. It is the *substantial law* of **LP**.

If a sequence of sets satisfies conditions (1) to (4), the substantial law of **LP** has a well-defined truth value with respect to this sequence. We can thus distinguish between actual and potential models of **LP**:

Definition 2.3 (Potential model of LP) x is a potential model of the lever principle ($x \in \mathbf{M}_p(\mathbf{LP})$) if and only if there exist D, n, t such that

(1) $x = \langle D, \mathbb{R}, n, t \rangle$
(2) D is a finite, non-empty set
(3) $n : D \to \mathbb{R}$
(4) $t : D \to \mathbb{R}$.

Definition 2.4 (Models of LP) x is a model of the lever principle ($x \in \mathbf{M}(\mathbf{LP})$) if and only if

(1) $x \in \mathbf{M}_p(\mathbf{LP})$
(2) $\sum_{y \in D} n(y) \cdot t(y) = 0.$

Structures that satisfy the substantial law of **T** are called *models of* **T**, in line with well established conventions in model theory. *Potential models* of **T**, by contrast, are structures that meet the formal conditions for satisfying the substantial law of **T** but not necessarily that law itself. These formal conditions are given by characterisations of structures, i.e., formulas that have exactly one occurrence of a relation symbol R_i.

The structuralist representation scheme has it that there is a one-to-one correspondence between substantial laws and theory-elements. Theory-elements are individuated by substantial laws, where there is some freedom of choice as to which axioms are grouped together to form the substantial law. In many cases, it is just one formal axiom that makes up a substantial law. In the above example, the lever principle (applied to the equilibrium case) individuates the theory-element **LP**.

We have claimed above that the structuralist representation scheme qualifies as a frame system in the sense of Minsky [87]. At this point, we can see that the definition of a model of a theory-element, as well as the definition of a potential model, introduces a frame concept. For, these definitions specify conditions that the valuations of the relation symbols $R_1 \ldots R_n$ in a sequence $\langle D_1, \ldots, D_k, R_1, \ldots, R_n \rangle$ must satisfy in order for this sequence to be a potential model and an actual model, respectively. Hence, we can say that the elements of a sequence $\langle D_1 \ldots D_k,\ A_1, \ldots, A_m,\ N_1 \ldots N_p,\ T_1 \ldots T_q \rangle$ qualify as valuations of slots in a Minsky-frame. The whole frame that corresponds to a theory-element **T** is composed of two elements: first, a sequence of slots for sets and, second, assignment conditions in the form of the definitions of potential and actual models of **T**. Further assignment conditions, those that concern the type of interpretation of the base sets, are made informally. So, the base sets $D_1, \ldots, D_k$ are characterised by describing their intended interpretation, but this characterisation is not part of the definitions of the potential and the actual models of **T**.

It is now even possible to make sense of Minsky's distinction between simple and complex conditions for slot assignments. The definition of the potential models of some **T** specifies simple type conditions that the relations $N_1 \ldots N_p, T_1 \ldots T_q$ must satisfy. The substantial law, by contrast, which is a component of the definition of the actual models of **T**, specifies a complex condition since it relates at least two relations to one another. The assignment conditions of potential models are complex as well, but only in the sense that they connect a relation to the base sets in which this relation is interpreted.

2.3.3 *Intended Applications*

In its first exposition by Sneed [104], structuralist theory of science emerged from a combination of the method of set-theoretic predicates with the Ramsey view of scientific theories. Consequently, it was referred to as the *emended Ramsey view*. This view became refined later on in *An Architectonic for Science* [18]; there it was simply termed *structuralism*. To understand the empiricist inheritance in structuralism, it is worth reconsidering briefly the original Ramsey view.

Essential to Ramsey's and Carnap's accounts of scientific language is the division of descriptive symbols into a set V_o of observational terms and another set V_t of theoretical terms. A scientific theory can thus be formulated in a language $L(V_o, V_t)$. (Henceforth, $L(V)$ designates the logical language based on the descriptive vocabulary V.) The division of the descriptive vocabulary gives rise to a related distinction between T- and C-axioms among the axioms of a scientific theory. The T-axioms contain only V_t symbols as descriptive ones, while the C-axioms contain both V_o and V_t symbols and thus establish a connection between these two subvocabularies. TC designates some conjunction of the T- and C-axioms. Let $n_1 \dots n_k$ be the elements of V_o and $t_1 \dots t_n$ the elements of V_t. Then, TC is a proposition of the following type:

$$TC(n_1 \dots n_k, t_1 \dots t_n). \qquad (TC)$$

The Ramsey sentence of a theory TC in the language $L(V_o, V_t)$ is obtained by the following two transformations of the conjunction of T- and C-axioms. First, replace all theoretical symbols in this conjunction by higher-order variables of appropriate type. Then, bind these variables by higher-order existential quantifiers. As result one obtains a higher-order sentence of the following form:

$$\exists X_1 \dots \exists X_n TC(n_1, \dots, n_k, X_1, \dots, X_n) \qquad (TC^R)$$

where $X_1, \dots, X_n$ are higher-order variables. If one thinks that TC^R is semantically more appropriate than representing a scientific theory by TC, one holds the *Ramsey view* of scientific theories.

In the original Ramsey view, empirical facts are represented by the interpretation of V_o, the observational vocabulary. $L(V_o)$ is a first-order language; higher-order logic may only be admitted for the representation of rational and real numbers. The representation of empirical facts in the emended Ramsey view is in a certain sense more structured than in the standard Ramsey view and standard first-order representations of scientific theories. In place of an interpretation of a first-order language $L(V_o)$, we have sequences of the type $\langle D_1 \dots D_k,\ A_1, \dots, A_m,\ N_1 \dots N_p \rangle$ that are used to represent a collection of empirical facts. By means of this style of representation, one aims to group together those empirical facts to which a certain theory-element **T** is applied. In the case of the above introduced lever principle, we have structures of type $\langle D, n \rangle$ to represent empirical phenomena, where D stands for the set of objects placed on the lever and n for the distance function.

It is an important feature of structuralist theory representation that a phenomenon is always associated with a theory-element that is intended to be applied to this phenomenon. To indicate that a structure y of type $\langle D_1 \dots D_k,\ A_1, \dots, A_m,\ N_1 \dots N_p \rangle$ represents a phenomenon to which a theory-element **T** is applied, one writes

$$y \in \mathbf{I}(\mathbf{T}).$$

I(T) designates the set of intended applications of the theory-element **T**. The structuralist framework, thus, forces one to represent single empirical facts as elements within a unit of theory application. This is comparable to the object-oriented style of programming forcing one to represent single data as members of a data structure that is associated with certain methods. Not surprisingly, the notion of an intended application can also be motivated by Minsky's request for inter-propositional knowledge representation, i.e., the association of facts with information about how to use these facts.

Consideration of certain important applications of classical mechanics will lead one to conclude that there are theories where it is even impossible to specify precisely, in a first-order setting, which empirical systems are phenomena of actual theory application. Assume we have a non-set-theoretic representation of the space-time and the mass function of all the minor and major objects of our solar system, including satellites. If no qualification is made saying to which subset of objects the law of gravitation is applied, we would have to consider every object of our solar system in order to determine the gravitational force acting upon a given object of that system. This, however, is not the way the laws of mechanics are used. Rather, certain subsystems, such as the interaction between Earth and Moon, are considered selectively for the application of the laws of mechanics in general and the law of gravitation in particular. Notably, the derivation of Kepler's laws from the axioms of Newton's mechanics is brought about by considering only two-body systems.

More generally, it can be observed that the selection of subsystems for the purpose of explanation and prediction is an important aspect of scientific reasoning and theory application. It is not obvious how this aspect could be expressed by first-order vocabulary (without set theory). Even if this should be feasible, set-theoretic means lead apparently to a more perspicuous representation.

We shall emphasise, finally, that the structuralist distinction between **T**-theoretical and **T**-non-theoretical concepts and symbols, even though reminiscent of the old empiricist distinction between theoretical and observational concepts, differs in important respects from that original distinction. The account of theoretical concepts in the structuralist framework emerged from the observation that, for every physical theory T, there are concepts whose meaning is determined through some axioms of **T**. This observation gives rise to the following informal explanation of a **T**-theoretical concept [18, p. 50]:

Explanation 2.1 "A concept whose determination involves some kind of measurement will be called theoretical with respect to theory T if all methods of measurement involved in its determination have to be conceived as models of T or as presupposing some models of T."

This explanation introduces a distinction between theoretical and non-theoretical concepts that is strictly relativised to a particular theory T. One and the same concept may be T_1-theoretical but not T_2-theoretical. For example, the concept of spatiotemporal position is theoretical with respect to certain theories about physical space-time, but not theoretical with respect to classical mechanics. Consequently, no assumption is made as to whether there is a level of observation that is free of any theory. Nor is any assumption made in structuralism as to whether we can acquire secure knowledge of **T**-non-theoretical propositions. Such propositions are not immune from revision. Whenever we distinguish between empirical facts and theoretical claims, this distinction is relative to a specific theory T or a specific theory-element **T**. This is an important refinement of Carnap's original double-language semantics in [33].

On the other hand, undeniably, there are empiricist elements that the structuralist approach inherits from Carnap and Ramsey. The empirical proposition associated with a theory-element **T** can, to some extent, be understood in analogy to the original Ramsey view. There, the theory's proposition was represented by the Ramsey sentence, i.e., by the claim that there is an interpretation of the theoretical terms such that TC becomes a true sentence in the context of a given valuation of the observation terms. In a similar vein, the proposition associated with a theory-element **T** has, as one component, the proposition that, for any intended application there is a valuation of the **T**-theoretical relation symbols such that the corresponding **T**-theoretical extensions are models of **T**. In more formal terms:

$$\forall y(y \in \mathbf{I}(\mathbf{T}) \rightarrow \exists x(x \in EXT(\mathbf{T})(y) \wedge x \in \mathbf{M}(\mathbf{T}))). \tag{2.5}$$

$x \in EXT(\mathbf{T})(y)$ says that x is a **T**-theoretical extension of y. To see the analogy between (2.5) and the Ramsey sentence more clearly, note that (2.5) translates to

$$\forall y(y \in \mathbf{I}(\mathbf{T}) \rightarrow \exists X_1, \ldots, X_q(x = \langle (y)_1, \ldots, (y)_{k+m+p}, X_1, \ldots, X_q \rangle \wedge \\ x \in \mathbf{M}(\mathbf{T})))$$

where $(y)_i$, $1 \leq i \leq k + p + m$, denotes the i-th element of the sequence y. $X_1, \ldots, X_q$ are higher-order variables that correspond to theoretical relations in the axioms of the theory-element **T**, just as the higher-order variables of the Ramsey sentence correspond to theoretical relations in TC. Hence, (2.5) must be read as implying the Ramsification of the substantial law of **T** for all of its intended applications. The whole empirical proposition associated with **T** is more complex than (2.5) since it also involves links to other theory-elements. These will be dealt with in the next section.

2.3.4 Links and Theory-Nets

Intended applications of theory-elements are not monadic items but rather related to one another in various ways. One distinguishes between three kinds of relations:

(1) Internal links
(2) External links
(3) Specialisation links.

These relations concern intended applications with their **T**-theoretical extensions. To be more precise, links further specify which theoretical extensions of an intended application are admissible, in addition to the requirement that any admissible theoretical extension must be a model of the respective theory-element. Let us begin with internal and external links before we come to deal with specialisations.

Elaborating upon the terminology of Minsky, we can say that links define *trans-frame* conditions in the sense that such conditions relate the slot assignments of different frame instantiations to one another. Among such conditions, we can distinguish between those that concern different instantiations of a single frame and others that concern instantiations of different frames. This, in essence, is the distinction between internal and external links: the former relate the intended applications of one and the same theory-element to one another, whereas the latter are binary relations among intended applications of different theory-elements.

The motivation for introducing internal and external links derives from overlapping intended applications. Two intended applications overlap if an only if there is an empirical object that occurs in some base set of both applications. For example, one and the same macroscopic object a may be placed several times on the lever of a balance, where it is combined with different sets of other empirical objects. Such is the use of the standard weights of a balance that works on the basis of the lever principle. In this case, we have different intended applications of the above introduced theory-element **LP**. The theoretical extensions of all overlapping intended applications must agree on the value they assign to the mass of a, which expresses the assumption that the mass of an object remains constant as long as only the object's spatiotemporal position is changed. This is an example of an internal link.

External links are particularly important to account for the transfer of data between intended applications of different theory-elements. Such a transfer obtains when **T**-non-theoretical relations of a theory-element **T** are determined with the help of a measuring theory $\mathbf{T}'$. The general motivation for introducing external links is similar to the motivation for internal ones: two intended applications of different theory-elements may overlap insofar as one and the same empirical object occurs as a member of the empirical base sets of those intended applications. External links are always binary.

Another type of relation among theory-elements is introduced to account for the inner structure of theories. A large number of scientific theories, in the ordinary sense of that term, have been shown to have a tree-like structure with a central theory-element at the top and several branches of more special theory-elements. The underlying idea is that any theory-element **T** is also an intended application of the more general theory-elements, which are higher up in the hierarchy. Hence, by means of specialisation, the substantial laws of different theory-elements can be superimposed. In formal terms, it holds:

$$\sigma(\mathbf{T}', \mathbf{T}) \rightarrow \forall x (x \in \mathbf{I}(\mathbf{T}') \rightarrow x \in \mathbf{I}(\mathbf{T})) \tag{2.6}$$

where $\sigma(\mathbf{T}', \mathbf{T})$ stands for the proposition that $\mathbf{T}'$ is a specialisation of $\mathbf{T}$. If $\sigma(\mathbf{T}', \mathbf{T})$, then it is a consequence of (2.5) and (2.6) that, for any intended application y of $\mathbf{T}'$, there must be a corresponding $\mathbf{T}$-theoretical extension x such that $x \in \mathbf{M}(\mathbf{T})$ and $x \in \mathbf{M}(\mathbf{T}')$. Hence, specialisations concern, just as internal and external links do, the interpretation of $\mathbf{T}$-theoretical extensions of intended applications.

A simple example of specialisation can be given with classical collision mechanics [88]. There, we have one central theory-element whose substantial law states the conservation of momentum:

Definition 2.5 (Potential models of CCM) x is a potential model of classical collision mechanics ($x \in \mathbf{M}_p(\mathbf{CCM})$) if and only if there exist P, T, v, m such that

(1) $x = \langle P, T, \mathbb{R}, v, m\rangle$
(2) P is a finite, non-empty set
(3) $T = \{t_1, t_2\}$
(4) $v : P \times T \to \mathbb{R}^3$
(5) $m : P \to \mathbb{R}_+$.

Definition 2.6 (Models of CCM) x is a model of collision mechanics ($x \in \mathbf{M}(\mathbf{CCM})$) if and only if there exist P, T, v, m such that

(1) $x = \langle P, T, \mathbb{R}, v, m\rangle$
(2) $x \in \mathbf{M}_p(\mathbf{CCM})$
(3) $\sum_{y \in P} m(y) \cdot v(y, t_1) = \sum_{y \in P} m(y) \cdot v(y, t_2)$.

This law has two specialisations, one for elastic and another for inelastic collisions. The conservation of kinetic energy is distinct of elastic collisions:

Definition 2.7 (Models of ECCM) x is a model of elastic classical collision mechanics ($x \in \mathbf{M}(\mathbf{ECCM})$) if and only if there exist P, T, v, m such that

(1) $x = \langle P, T, \mathbb{R}, v, m\rangle$
(2) $x \in \mathbf{M}_p(\mathbf{CCM})$
(3) $\sum_{y \in P} m(y) \cdot |v(y, t_1)|^2 = \sum_{y \in P} m(y) \cdot |v(y, t_2)|^2$.

Inelastic collisions, by contrast, violate the conservation of kinetic energy. Equality of velocity of all colliding particles results from an inelastic collision:

Definition 2.8 (Models of ICCM) x is a model of inelastic classical collision mechanics ($x \in \mathbf{M}(\mathbf{ICCM})$) if and only if there exist P, T, v, m such that

(1) $x = \langle P, T, \mathbb{R}, v, m\rangle$
(2) $x \in \mathbf{M}_p(\mathbf{CCM})$
(3) for all $y, z \in P : v(y, t_2) = v(z, t_2)$.

Collision mechanics is thus found to have a simple tree structure presented by the following figure (Fig. 2.2):

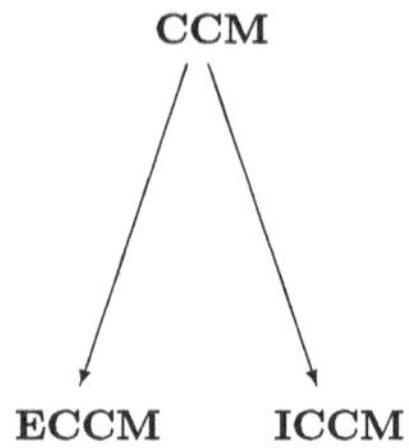

Fig. 2.2 Specialisations of CCM

Arguably, one can also view specialisation as a subspecies of external links. For, a specialisation relates two intended applications of different theory-elements with one another and, thereby, constrains the actual models of the more specific theory-element.

A fully fledged scientific theory, such as classical mechanics or phenomenological thermodynamics, normally consists of a number of theory-elements related to one another by internal, external and specialisation links. Such units of inter-related theory-elements are called *theory-nets*. The notion of a theory-net, however, is not confined to a single scientific theory but may well comprise the theory-elements of a number of such theories – in the ordinary sense of the term *theory*.

The notion of a theory-holon is even more expressive than that of a theory-net. This notion takes such inter-theoretical relations as reduction and equivalence among fully fledged scientific theories into account. For the present investigation, however, theory-nets are sufficiently expressive. Theory-nets, therefore, figure as units for which revisions of beliefs are defined. We shall write $\mathbf{T} \in \mathbf{N}$ to indicate that the theory-element **T** is a member of the theory-net **N**. Formally, a theory-net **N** is a set of theory-elements.

In essence, structuralist theory representation consists in specifying which theoretical extensions of a given set of intended applications are admissible. The application of scientific theories to empirical phenomena can be perceived, consequently, as a *constraint satisfaction problem*, quite in the computer scientist's understanding of such problems [97]. Notably, the global empirical claim of a theory-net **N** consists in the statement that a certain constraint satisfaction problem has a solution. This problem is to find a set $A(\mathbf{N})$ of structures such that (i) for all $y \in I(\mathbf{T})$ $(\mathbf{T} \in \mathbf{N})$, there is an $x \in A(\mathbf{N})$ such that x is a model of **T** and a theoretical extension of y, and (ii) all members of $A(\mathbf{N})$ satisfy all internal, external and specialisation links of **N**. In Chap. 5, a precise formulation of this global empirical claim will be given.

Chapter 3
Belief Revision

This chapter introduces the classical AGM approach to belief revision as well as the basic ideas of belief base revision theory. Finally, we describe a number of open problems that will be tackled in Part II of the investigation.

3.1 Classical Belief Revision Theory

3.1.1 Beliefs

The notion of a belief does not seem to allow for a reductive explanation. It is feasible, though, to relate the notion of a belief to other notions in an informative way. So, we can say that beliefs form a particular type of propositional attitude. But what is a proposition? We shall be content here with saying that propositions are represented by meaningful sentences. Purely extensional accounts of propositions in terms of possible worlds are thus excluded. The notion of meaning is understood along the lines of the Fregean notion of sense and an explanation thereof by Church [38, p. 6]:

> Of the sense we say that it determines the denotation, or is a concept of the denotation.

Implicit in this explanation is that it is syntactic entities that have a sense. For, denotation is a relation between syntactic and semantic entities. Applied to sentences, Church's explanation says that the sense of a sentence is what determines its truth value.

The affirmation, or acceptance, of a meaningful sentence α is what is called the *belief that* α. Correspondingly, the negation, or rejection, of a meaningful sentence α may be called the *belief that* $\neg\alpha$. Hence, the negation of α amounts to the acceptance of $\neg\alpha$. Both, affirmation and negation are attitudes toward meaningful sentences and because of this also attitudes toward propositions. A third attitude that a speaker can

H. Andreas, *Dynamic Tractable Reasoning*, Synthese Library 420,
https://doi.org/10.1007/978-3-030-36233-1_3

adopt is the neutral one, which consists in neither believing nor disbelieving α. Her attitude toward α is then indeterminate. Acceptance, rejection and the indeterminate attitude toward a proposition are epistemic attitudes. Any epistemic attitude toward a proposition – be it affirmative, negative or indeterminate – is one of a particular speaker or a group of speakers. The reference to a speaker or a group of speakers remains, however, implicit in most investigations of classical belief revision theory.

The sentential representation of propositions is fundamental to the methods of belief revision theory simply because this theory assumes such a representation [61, 69, 98]. More precisely, beliefs are represented by sentences of a formalised logical language. Such languages may be interpreted directly in the way suggested by model-theoretic semantics of predicate logic. Another way of interpreting a formal language is to give designation rules for the descriptive symbols, as originally suggested by Carnap [32]. Here is a simple example:

$\ulcorner m(x) \urcorner$ designates the mass function.

Such designation rules relate the descriptive symbols of the formal language to expressions of a non-formalised language and thus interpret symbols. This type of interpretation is best seen as complementary to the extensional interpretation of model-theoretic semantics. It is very common to specify the intended interpretation of a formalised language by means of Carnapian designation rules.

How is the notion of truth related to that of acceptance? On the one hand, it is certainly correct to say that our policy of accepting sentences aims at true sentences. On the other hand, it is not plainly obvious that the assignment of truth values is, for any type of sentence, settled in advance to and independently of our acceptance and rejection of sentences. The spectrum of epistemological and metaphysical doctrines has two interesting limiting cases. At the one end, truth is seen in strictly non-epistemic terms, with the consequence that truth values are settled independently of our methods of determining such values. At the other end, one may even attempt to dispense with truth values altogether and reinterpret these values as values for the attitudes of acceptance and rejection. This strategy is particularly tempting in view of truth-value semantics, where first- and higher-order languages are interpreted by truth-value assignments to atomic sentences [78]. Of course, if truth values are replaced with values for acceptance and rejection, the interpretation of the language will remain, most of the time, a partial one in the sense that some sentences do not have a determinate value. The extreme form of an antirealist semantics is encapsulated in the thought that there is no difference between knowledge and full belief.

Between these two limiting cases, various positions are located, according to which there is a certain range of sentences whose truth values are determined, at least in part, by our acceptance or rejection of these or other sentences. Interesting candidates for sentences with that property are Poincarean conventions and Carnapian postulates. In the course of the present investigation, we will touch upon epistemological issues from time to time. Whenever possible, however, we shall avoid explicit commitment to epistemological and metaphysical doctrines. As for the relation of truth to acceptance, a determinate view need not necessarily be adopted to investigate the dynamics of belief. This has been demonstrated in

Gärdenfors's seminal book on belief revision theory, in which the concept of truth proved irrelevant for the investigation of rationality criteria governing belief changes [61, p. 9].

3.1.2 *Belief Sets*

Belief revision theory is, to a large extent, a study of the rationality criteria that are presumed to govern our beliefs. Among such rationality criteria there are those that concern the static dimension of beliefs and others that concern their dynamic dimension. An important static rationality criterion is the requirement that one should believe all the logical consequences of one's explicit beliefs. In more technical terms, epistemic states are represented by sets of sentences that are closed under the operation Cn of logical consequence. A set A of sentences is only a belief set if it holds that $Cn(A) = A$. Cn is defined by the semantics of the underlying logic. $A \models \alpha$ is an alternative notation for $\alpha \in Cn(A)$.

Which logics are admitted to underlie our inferential practice in classical belief revision theory? This question is answered by imposing some constraints upon the operation Cn of logical consequence. Hansson [69, p. 26] distinguishes between the definitional and additional properties that Cn is assumed to satisfy. Let us begin with the definitional ones:

(1) $A \subseteq Cn(A)$. (Inclusion)
(2) If $A \subseteq B$, then $Cn(A) \subseteq Cn(B)$. (Monotony)
(3) $Cn(A) = Cn(Cn(A))$. (Iteration)

Requiring monotonicity of Cn is quite a severe restriction in view of the developments nonmonotonic logics. However, requiring that a belief set A is logically closed under a monotonic consequence operation does not imply that all derived items of A must be acquired by means of classical monotonic reasoning. Inclusion and iteration also hold in standard nonmonotonic logics.

In addition to the definitional properties, Cn is assumed to meet some further requirements [69, p. 26]:

(1) If α can be derived from A by classical truth-functional logic, then $\alpha \in Cn(A)$. (Supraclassicality)
(2) $\beta \in Cn(A \cup \{\alpha\})$ if and only if $(\alpha \rightarrow \beta) \in Cn(A)$. (Deduction)
(3) If $\alpha \in Cn(A)$, then $\alpha \in Cn(A')$ for some finite subset $A' \subseteq A$. (Compactness)

According to the first of these properties, Cn contains classical truth-functional consequence. It is thus assumed that the underlying logic contains the standard truth-functional connectives for negation, conjunction, disjunction, implication and equivalence. Classical belief revision theory is rarely concerned with the sub-propositional level so that relations, functions, and quantifiers play no significant role in most investigations.

It is uncontroversial that the identification of epistemic states with belief sets is an idealisation. In practice, it holds only for a minor fraction of the items of a belief set that the sentence is a belief in the sense of believing with awareness. A particular problem is posed by the fact that any belief set contains infinitely many beliefs that seem to be of no interest and completely uninformative. To give an example, for any explicit belief α and any arbitrary well-formed sentence β of L, the belief set will contain the sentence $\alpha \vee \beta$ regardless of the epistemic attitude toward β. A more realistic model has been advanced by the study of *belief bases*, which will be dealt with in Sect. 3.2.

One word on the notion of an epistemic state is in order. Because of the variety of approaches to belief revision, there is no unique explanation or definition of this notion available. Gärdenfors [61] and Hansson [69] model epistemic states by belief sets. In semantic approaches to belief revision, epistemic states are modelled by an ordering of possible worlds (see, e.g., Ditmarsch [46] and Spohn [105]). As we shall see below, the revision of a belief set K by a sentence α is well defined only if we assume some epistemic ordering on K or an ordering on a set related to K, such as a belief base H of K. This suggests understanding the notion of an epistemic state – in syntactic approaches to belief revision – such that this notion includes information about an epistemic ordering. We shall adopt this understanding when we come to investigate means of tractable belief revision.

3.1.3 AGM Postulates for Belief Changes

The study of belief changes in the original AGM approach is concerned with transformations of belief sets. Three types of changes of a belief set K are distinguished:

(1) Expansion with a sentence α $(K + \alpha)$
(2) Revision with a sentence α $(K * \alpha)$
(3) Contraction by a sentence α. $(K \div \alpha)$

Expansions are intended to capture the case where a proposition α, which is presumed to be consistent with the belief set K, is added to K. We speak of a revision of K by α, by contrast, if α is possibly inconsistent with the belief set K. The revision of K by α must result in a consistent belief set K' that contains α. The contraction by a proposition α, finally, is the transformation that K undergoes when an element α of K is eliminated.

The AGM framework evolved in the form of two interrelated approaches. First, belief changes have been characterised by means of rationality postulates, which amounts to an axiomatic approach to belief revision. Second, in the constructive approach, explicit set-theoretic operations on belief sets have been introduced to model belief changes. These two approaches could be related to one another by several representation theorems. Such theorems state that certain postulates are satisfied by a belief change if and only if this belief change is achieved by a sequence of certain well-defined set-theoretic operations on the belief set.

3.1.3.1 Expansions

Expanding a belief set K by a proposition α is an operation that is rather easy to characterise using axioms. It allows, moreover, for a straightforward and unique explicit definition:

$$K + \alpha = Cn(K \cup \{\alpha\}). \qquad \text{(Def +)}$$

The expansion $+$ of a belief set K with α is obtained by first adding α to K and, second, the operation of logical closure. Expansions do not pose difficult problems because this type of operation is intended to apply to cases where the epistemic input is consistent with the belief set K. No revision of old beliefs is therefore required to integrate α into K. In case α is inconsistent with K, applying (Def +) leads to $K+\alpha = L$, i.e., the belief set that contains every sentence of the language L. This set is designated by $K_\perp$ and called the *absurd*, or the *contradictory*, belief set. Because of its relative simplicity, the axiomatic characterisation of expansions dropped out of more recent accounts of belief revision. The above explicit definition, however, has remained important and will be referred to in the sequel.

More intricate than expansions are revisions and contractions. For these operations there is no simple and straightforward definition in terms of set-theory and logic. Nor do rationality postulates alone suffice to define contractions and revisions uniquely, for reasons that will become obvious soon. It proved therefore necessary to assume some sort of epistemic ordering among the items of a belief set in order to define revisions and contractions of such a set in a unique manner. In what follows, we shall give a brief account of the postulates for revisions and contractions, and then move on to the constructive approach.

3.1.3.2 Revisions

(K^*1)	$K * \alpha$ is a belief set.	(Closure)
(K^*2)	$\alpha \in K * \alpha$.	(Success)
(K^*3)	$K * \alpha \subseteq K + \alpha$.	(Expansion 1)
(K^*4)	If $\neg\alpha \notin K$, then $K + \alpha \subseteq K * \alpha$.	(Expansion 2)
(K^*5)	$K * \alpha = K_\perp$ only if $\neg\alpha \in Cn(\emptyset)$.	(Consistency preservation)
(K^*6)	If $\alpha \leftrightarrow \beta \in Cn(\emptyset)$, then $K * \alpha = K * \beta$.	(Extensionality)

These postulates are commonly referred to as the *basic Gärdenfors postulates for revision*. Two supplementary postulates concern the revision of a belief set by a conjunction:

(K^*7)	$K * \alpha \wedge \beta \subseteq (K * \alpha) + \beta.$	(Conjunction 1)
(K^*8)	If $\neg\beta \notin K * \alpha$, then $(K * \alpha) + \beta \subseteq K * \alpha \wedge \beta.$	(Conjunction 2)

3.1.3.3 Contractions

$(K^{\div}1)$	$K \div \alpha$ is a belief set.	(Closure)
$(K^{\div}2)$	$K \div \alpha \subseteq K.$	(Inclusion)
$(K^{\div}3)$	If $\alpha \notin K$, then $K \div \alpha = K.$	(Vacuity)
$(K^{\div}4)$	If $\alpha \notin Cn(\emptyset)$, then $\alpha \notin K \div \alpha.$	(Success)
$(K^{\div}5)$	$K \subseteq (K \div \alpha) + \alpha.$	(Recovery)
$(K^{\div}6)$	If $\alpha \leftrightarrow \beta \in Cn(\emptyset)$, then $K \div \alpha = K \div \beta.$	(Extensionality)

The postulates $(K^{\div}1)$–$(K^{\div}6)$ are commonly referred to as the *basic Gärdenfors postulates for contractions*. The two supplementary postulates for revisions, (K^*7) and (K^*8), have close relatives for contractions:

$(K^{\div}7)$	$K \div \alpha \cap K \div \beta \subset K \div \alpha \wedge \beta.$	(Conjunction 1)
$(K^{\div}8)$	If $\alpha \notin K \div \alpha \wedge \beta$, then $K \div \alpha \wedge \beta \subseteq K \div \alpha.$	(Conjunction 2)

It is worth noting that contractions and revisions are interdefinable. Particularly intuitive is what has come to be called the *Levi identity*: the revision of K by α is equivalent to a contraction of K by $\neg\alpha$ and a subsequent expansion by α:

$$K * \alpha = (K \div \neg\alpha) + \alpha. \qquad \text{(Def } *\text{)}$$

Not quite as intuitive as the Levi identity is the other direction, from revisions to contractions, which is expressed by the *Harper identity*:

$$K \div \alpha = (K * \neg\alpha) \cap K. \qquad \text{(Def } \div\text{)}$$

These two identities receive strong support through representation theorems showing that, if one of the two operations satisfies certain fundamental rationality postulates, then the other operation, which is defined by either the Levi or the Harper identity, will satisfy its fundamental rationality postulates as well [61, pp. 68–71].

3.1.4 Constructive Approaches

Various constructive approaches to revisions and contractions have been devised since the seminal AGM paper appeared [1]. We shall focus here on Gärdenfors's [61] entrenchment-based revisions. Let us begin with the formal characterisation of the epistemic entrenchment relation. $\alpha \leq \beta$ is to be read as 'α is at most as epistemologically entrenched as β'. The following postulates formally characterise this relation [61, pp. 89–91]:

(EE 1) If $\alpha \leq \beta$ and $\beta \leq \chi$, then $\alpha \leq \chi$. (Transitivity)
(EE 2) If $\alpha \models \beta$, then $\alpha \leq \beta$. (Dominance)
(EE 3) $\alpha \leq \alpha \wedge \beta$ or $\beta \leq \alpha \wedge \beta$. (Conjunctiveness)
(EE 4) When $K \neq K_{\perp}$, $\alpha \notin K$ iff $\alpha \leq \beta$ for all $\beta \in L$. (Minimality)
(EE 5) If $\beta \leq \alpha$ for all β, then $\alpha \in Cn(\emptyset)$. (Maximality)

Epistemic entrenchment orderings and contractions are interdefinable. Assume a contraction function $\div$ is given. Then, an epistemic entrenchment ordering can be introduced via the following definition [69, p. 100]:

(C $\leq$) $\alpha \leq \beta$ iff $\alpha \notin K \div (\alpha \wedge \beta)$ or $(\alpha \wedge \beta) \in Cn(\emptyset)$.

If a contraction with $\alpha \wedge \beta$ leads to the elimination of α, then α cannot be more epistemically entrenched than β. Otherwise, it would be retained in the contracted belief set. This is the rationale of the first part of the definiens. In case $\alpha \wedge \beta$ is a logical truth, then both α and β are logical truths, which would imply that they have equal epistemic entrenchment. Then $\alpha \leq \beta$ holds as well. This is the second part of the definiens.

The other direction, from epistemic entrenchment orderings to contractions, provides the second constructive approach to belief revision [69, p. 100]:

(G $\div$) $\beta \in K \div \alpha$ iff $\beta \in K$ and either $\alpha < (\alpha \vee \beta)$ or $\alpha \in Cn(\emptyset)$.

So, a belief β of K will remain in the belief set after a contraction with α if and only if either α is (strictly) less epistemically entrenched than $\alpha \vee \beta$ or α is a logical truth.[1] As one would expect it, the relation $<$ of being strictly less epistemically entrenched is defined such that $\alpha < \beta$ if and only if $\alpha \leq \beta$ but not $\beta \leq \alpha$. Once contractions are defined, revisions can be determined using the Levi identity (cf. Sect. 3.1.3).

The importance of an epistemic entrenchment ordering derives from that it allows for a unique determination of the contraction function for a given belief set K. The rationality postulates, by contrast, define only classes of such functions. A unique determination of contractions and revisions is desirable since it is reasonable to require that the revision of a belief set K by α leads to a determinate new belief set K'. This requirement is formally expressed by the postulates $(K^{\div}1)$ and $(K^{*}2)$.

What determines the epistemic entrenchment ordering for a given belief set? What conceptual explanation can be adduced of such an ordering? It is obvious that the postulates (EE1)–(EE5) leave room, apart from very special cases, for numerous extensional interpretations of the epistemic entrenchment ordering. Hence, (EE1)–(EE5) are insufficient to determine the epistemic entrenchment ordering. Nor would it be fully appropriate to say that the epistemic entrenchment orderings

[1] The justification of the condition $\alpha < (\alpha \vee \beta)$ is somewhat lengthy and cannot be encapsulated in a brief argument. For this the reader is referred to Hansson [69, p. 100].

are empirically determined by factually occurring revision processes together with (C<). Gärdenfors's view was rather that epistemic entrenchment orderings are determinable independently of such processes [61, p. 87] (emphasis not added):

> It is possible to determine the relative epistemic entrenchment of the sentences in a belief set K *independently* of what happens to K in contractions and revisions.

Having clarified this, he goes on to explain what truly determines the ordering of beliefs [61, p. 87]:

> The fundamental criterion for determining the epistemic entrenchment of a sentence is how useful it is in inquiry and deliberation.

From these statements we can infer that the epistemic entrenchment ordering is expected to explain contractions and revisions as they actually happen to our epistemic attitudes.

It continues to be an open question, however, how the usefulness of sentences in inquiry and deliberation can be determined. Gärdenfors himself cites the information-theoretic approach, according to which epistemic entrenchment could be measured in terms of the information that would be lost if the sentence were to be retracted from the belief set [61, p. 92]. At the same time, he sees several problems with measuring information loss in the context of belief revision. Furthermore, metatheoretical structuralism in the form of Sneed [104] and Stegmüller [106] is referred to as a means to establish a hierarchy among the axioms of scientific theories that reflects their epistemic entrenchment. This reference is of great interest in the context of our investigation since it lends some support to the conjecture that structuralist theory of science may fruitfully be connected with belief revision theory. We shall further discuss the problem of determining an epistemic entrenchment ordering in Sect. 3.3.

3.2 Belief Base Revision

3.2.1 Belief Bases

The study of belief bases and revisions thereof is intended to achieve a more realistic representation of epistemic states and their dynamics. It can be seen, therefore, as an emendation of classical belief revision theory, which has been concerned with changes of (logically closed) belief sets. Why are belief sets felt to be a deficient representation of epistemic states from a cognitive point of view? The problem is that even for languages of propositional logic, any belief set is infinite. This contrasts with the finiteness of human minds and computers. As human minds have only a finite capacity to memorise sentences that are accepted, so have computers only a finite storage.

Unlike belief sets, belief bases are allowed to be finite and are usually assumed to be so. The idea is to have a set H of explicit beliefs that represents all further implicit beliefs in the sense that the latter beliefs are consequences of H. This gives rise to the following formal definition of a belief base (cf. [69, p. 18]):

Definition 3.1 (Belief base) Any set H of sentences is a belief base. Let K be a belief set. Then, a set H of sentences is a belief base for K if and only if there is a monotonic consequence operation Cn or a nonmonotonic inference operation Inf such that $K = Cn(H)$ and $K = Inf(H)$, respectively.

Working with a nonmonotonic inference operation allows to account for a wider range of inferential patterns that justify the items of the belief set on grounds of the items of the belief base [24]. This strategy will be adopted for the present investigation.

In addition to $K = Cn(H)$ or $K = Inf(H)$, another property is considered essential for the members of a belief base. These members may be non-derived in the sense that they have been acquired non-inferentially. Rott gives the following explanation of a belief base [98, p. 23]:

> The elements of a base are explicitly given beliefs. They comprise beliefs and only beliefs, which have some kind of independent standing, i.e., which are not derived by processes of inference from other beliefs. They serve as a basis or foundation of the belief set that can be 'built on' the base.

Note, finally, that truth is not a requirement for the elements of a belief base.

3.2.2 Belief Bases vs. Belief Sets

Why study belief changes via belief bases rather than belief sets? One important reason for this choice is the above indicated finite-memory argument for human minds and computers. Of equal importance seems to be the fact that belief base revisions respect justifications more properly than existing formalisms for belief set revisions. A belief set is not an entity that can be used to express that certain beliefs are reasons for other beliefs [61, p. 67].

The failure to respect justifications may lead to a counter-intuitive behaviour of belief sets when it comes to the retraction of beliefs that serve as justifications for other beliefs. Assume $E, \alpha \models \beta$, where E is some background theory and $E \subset K$. Further, $\alpha, \beta \notin K$ and $Cn(K \cup \{\beta\}) \neq K_{\perp}$, and $E \subset K * \alpha$. Then, β is a member of $K * \alpha$. Now, if we contract K by α, β may well be retained in the resulting belief set, even though the sole justification of β has been lost. Indeed, according to a strict reading of the maxim of minimum mutilation, β must be retained. This is a well-known problem in classical belief revision theory ([61, p. 67] and [70, p. 6]). It amounts to a violation of the following intuitive reversion principle:

$$(A * \alpha) \div \alpha \subseteq A. \qquad \text{(Rev 2)}$$

This principle can be seen as the counterpart of the famous, yet controversial recovery postulate ($K\dot{-}5$). While recovery requires that any belief lost through contraction with α may be recovered through addition of α, the above reversion principle requires that any belief gained through revision with α should be dismissed if α is contracted afterwards.

It is obvious that a corresponding problem does not arise in belief base revision. If a sentence α is added to the base and retracted afterwards, then all derived beliefs whose justification depends on α become retracted as well. Note, however, that the problem under consideration can be solved in the belief set approach to belief revision, viz., by adopting a postulate like (Rev 2).

With some qualifications, it holds that the study of base revisions is driven by a foundationalist epistemology, whereas belief set revision strategies are guided by a coherentist picture of justification [98, Chap. 2n]. Hence, the common repudiation of foundationalism in mainstream epistemology may be taken as an argument in favour of belief set revision strategies (or in favour of base revisions if one does not wish to follow the mainstream). However, the distinction between the belief base in the narrow sense and a corpus of background theories, which we will encounter in more elaborated variants of base revision strategies below, does not fit so well into textbook explanations of foundationalism. For, the elements of background theories need not and do not have a secure basis of justification. Nor is it required that the elements of the belief base in the narrow sense are true or self-evident. Research in belief revision may thus lead to a more viable account of foundationalist epistemology that is not so easy to refute as standard accounts discussed in the literature.

Hansson makes a subtle distinction between base revisions and base-generated revisions [69, pp. 305–321]. In the study of base revision, the epistemic state is identified with a determinate belief base. In base-generated revision, by contrast, epistemic states remain identified with belief sets. Any change operation of the belief set, however, must be representable as the outcome of the change of a corresponding belief base. The belief base thus works as a hidden structure of the epistemic state. The central motivation for studying base-generated revisions as opposed to base revisions is the *principle of the irrelevance of syntax* [42], which is sometimes violated in the belief base approach. For, when revising belief bases, it may happen that two logically equivalent bases behave differently. Compare $\{p \wedge q\} * \neg q$ with $\{p, q\} * \neg q$, where p and q have the same epistemic standing in $\{p, q\}$. Only in case of the latter revision does p remain accepted despite the logical equivalence between $\{p \wedge q\}$ and $\{p, q\}$.

Contrary to the principle that syntax be irrelevant, Brewka [24] has pointed out that choosing the formulation $\{p \wedge q\}$ over $\{p, q\}$ may well be intended to make a difference. This choice is justified, for example, if p and q are only to be given up together. Generally, one may argue that the operation of logical consequence does not capture the dynamic behaviour of sentences. Logical equivalence among belief bases, therefore, does not imply equivalent dynamic properties. We shall not, in the present investigation, attempt to respect the principle of the irrelevance of syntax. For this reason, no distinction is made between base revisions and base-generated revisions. A belief base H is thus required to contain only explicit and non-derived beliefs. It is not merely a compact representation of a corresponding belief set.

3.2.3 The Direct Mode of Belief Base Revision

For our investigation it is instructive to familiarise ourselves with Hans Rott's account of the *direct mode* of belief base revision. In this mode, the belief set K is generated from the base H by a nonmonotonic and paraconsistent inference operation. The idea is that changes of the belief base are made in a plain (artless, or unsophisticated) manner, as Hans Rott puts it. A revision by a sentence α simply consists in adding α to the base H, while a retraction by α is performed through subtracting α from it. Changes of the belief set K result from base changes and the underlying logic. Figure 3.1 gives a graphical representation of the direct mode [98, p. 84].

As just explained, the formation of beliefs is governed by a nonmonotonic and paraconsistent inference operation:

$$K = Inf(\mathcal{H}).$$

Paraconsistent and nonmonotonic systems that are used to define belief base revisions commonly assume an epistemic ranking among the items of the belief base:

$$\mathcal{H} = \langle H_1, \ldots, H_n \rangle.$$

Trivially, any basic belief must be an element of some H_i, $1 \leq i \leq n$. Following [98, p. 35], we assume that the sets $H_1, \ldots, H_n$ are disjoint. These sequences, then, represent a strict weak ordering among the items of the base, with the understanding that $\alpha \in H_i$ has priority over $\beta \in H_j$ just in case $i < j$. α's having priority over β – in symbols: $\alpha < \beta$ – means that α is more certain, more important, more valuable or more reliable than β.

$<$ is a strict weak ordering if and only of it is binary, transitive, asymmetric, and the incomparability relation $I(\alpha, \beta) \leftrightarrow \neg(\alpha < \beta) \wedge \neg(\beta < \alpha)$ is transitive. Such an ordering is also called *modular*. The sets $H_1, \ldots, H_n$ are equivalence classes

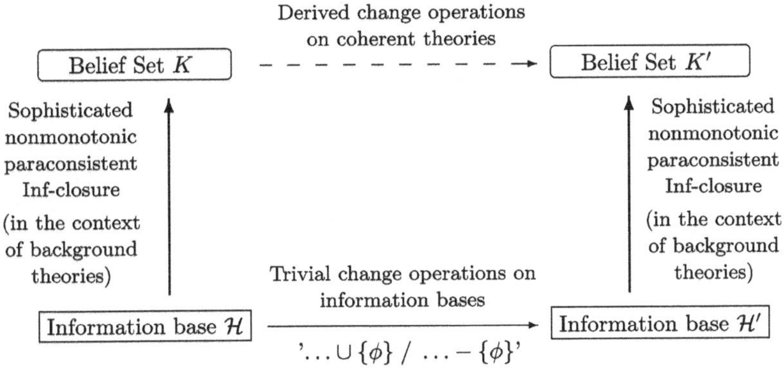

Fig. 3.1 The direct mode

with $I(\alpha, \beta)$ as equivalence relation. All members of a certain class are considered to have equal epistemic priority. So, having equal epistemic priority is defined by means of being incomparable according to the ordering relation $<$.

We call the sequence $\mathcal{H}$ a *ranked belief base* and an ordering relation $<$ on a belief base H an epistemic ranking. While $\mathcal{H}$ is sometimes also referred to as a *prioritised belief base*, we prefer to speak of a ranked belief base so as to indicate that the ordering relation is a strict weak ordering. More precisely, the epistemic ordering is a ranking that allows for ties. The notion of an epistemic priority ordering is more general since such an ordering may well be partial.

Once an appropriate inference operation Inf applying to $\mathcal{H}$ is at hand, the revision of a belief base by a sentence α can be defined in a straightforward manner [24]:

$$K *_i \alpha = Inf(\mathcal{H} \oplus_i \alpha).$$

$\mathcal{H} \oplus_i \alpha$ stands for expanding the set H_i by α. This means, in less technical terms, that α is inserted into the set H_i of the ranked base $\mathcal{H}$. One thus assumes that the epistemic input α has a certain level of epistemic priority, which results in a corresponding revision operation $*_i$. An alternative to this scheme is to assign new epistemic input top priority such that the revision formula has priority over all items of the old belief base $\mathcal{H}$ [98, p. 28]. This move guarantees satisfaction of the success postulate (K^*2).

A genuine virtue of the present approach to belief revision is that iterated belief changes are well defined right away. That is, belief revisions of the form $(K * \alpha) * \beta$ do not require additional elements. For most other belief set and belief base revision schemes in the AGM tradition, by contrast, the definition of iterated belief changes is rather challenging and not simple.

There is a simple and straightforward way of defining contractions [98, p. 131]:

$$K \div \alpha = Inf(\mathcal{H} \setminus \{\alpha\}).$$

$(\mathcal{H} \setminus \langle\alpha\rangle)$ is the operation of removing all occurrences of α in $\mathcal{H}$. This construction, however, may fail to remove α from the belief set K since α may well be a member of $Inf(\mathcal{H})$ without being a member of any H_i of $\mathcal{H}$. Brewka [24] has given a refined definition that guarantees elimination of the contraction formula from the belief set K.[2]

Rott [98, Chap. 5] distinguishes between a belief base in the narrow sense and a set E of *expectations* which are considered to belong to the belief base in the wider sense. Expectations are defeasible axioms of background theories that contribute to the generation of derived beliefs from the base. The defeasible nature of these axioms gives rise to speaking of expectations. This terminology is inspired by Gärdenfors and Makinson [62]. Notably, we may have an epistemic ranking among expectations:

[2]This definition makes use of constraints in the sense of Poole [92]. These are sentences that the belief set needs to be consistent with but which may not be used to derive new information.

$$\mathcal{E} = \langle E_1, \ldots, E_n \rangle.$$

The distinction between a belief base H (in the narrow sense) and a set E of expectations will prove fruitful in our approach. Either the basic beliefs or the expectations are to be ranked but not both.

To distinguish between beliefs that are basic in the narrow sense and expectations, we shall also speak of proper basic beliefs when intending to refer to the former. The idea is that proper basic beliefs are non-inferentially acquired, while expectations may be acquired by way of inductive generalisations or abductive inferences of theory formation.

The union of basic beliefs and expectations need not be classically consistent. The set of beliefs generated from $\mathcal{H}$ and $\mathcal{E}$, by contrast, better be consistent. Hence, it is the job of the logical machinery to remove inconsistencies in the union of basic beliefs and expectations. How this can be done will be explained in the next chapter.

If there is a direct mode of belief base revision, there of course is also an indirect mode. The latter uses the Levi identity so that the revision of a belief base is defined on the basis of its contraction. The contraction of a belief base H by α, in turn, is defined using maximal subsets of H that do not imply α. This definition assumes an epistemic priority ordering among the subsets of H. In what follows, we shall work with the direct mode of belief base revision, simply because we found solutions to the key problems of our investigation for this mode. For simplicity, we do not go into the details of the indirect mode.[3]

3.3 Some Open Problems in Belief Revision Theory

In his "Ten Philosophical Problems in Belief Revision Theory", Hansson [70] outlines problems for future research in belief revision. Our approach to belief revision aims to solve two of these ten problems and another, additional problem. In what follows, we describe the three problems in belief revision that are directly addressed in our investigation.

3.3.1 Tractability

At the present state of the art in belief revision theory, belief changes are computationally intractable. (We shall explain the precise meaning of intractability and the reasons why extant belief revision schemes are intractable in Sect. 7.1.) Hence, from a cognitive science perspective, belief changes must be considered cognitively infeasible. This is of course a paradoxical conclusion. The primary objective of

[3]The indirect mode of belief base revision originates with Hansson [68].

using frame concepts in an analysis of belief changes is to show how we can revise our beliefs in a tractable manner.

The relevance of tractability in cognitive science derives from CTM, the computational theory of the mind and the Church-Turing thesis – in ways explained in the Introduction. Notably, the Church-Turing thesis alone implies that the biological machinery supposedly underlying our faculty of reasoning can be simulated by a Turing machine. From this it follows that we can use the concepts of computational complexity theory in an analysis of the computational limitations of the human mind. Hansson [70] does not directly address the problem of tractability, but raises a closely related question: "Can stricter cognitive limitations than finiteness be represented in an interesting way?" (p. 38). As is obvious from the Introduction, the problem of tractability is the primary motivation for the present investigation.

3.3.2 The Justificatory Structure

Any conception of an epistemic ordering is closely related to an individual's resistance to change beliefs: the better the standing of a belief is in the ordering, the more will the individual resist retracting it. One would expect, therefore, that the epistemic standing of a belief has something to do with the justification that can be given for it. For, it seems uncontroversial that we particularly resist giving up beliefs that are well justified. In the study of belief set revisions, however, justifications play no apparent role in the determination of the epistemic ordering. The relation between the available justifications for a belief and its epistemic standing remained largely unexplored. Relatedly, Hansson [70, p. 41] raises the following question: "What is the relation between vulnerability/resistance and justificatory structure?"

The situation is much better for belief base revision. There, every derived belief is justified by a subset of the belief base. In the case of ranked belief bases it holds that the higher the rank is of those items which serve as justifications for a belief α, the more will the individual resist retracting α. As for the epistemic ordering of the belief base, Brewka [23] suggests that it be interpreted in terms of the reliability of basic beliefs. This idea will be further developed in the course of the present investigation.

There is a subtle but important difference between representing and respecting justifications. Belief base changes in the direct mode respect the justificatory structure of a belief system insofar as well-justified (derived) beliefs are less readily given up than those that are less well justified. Derived beliefs whose sole justification has been retracted from the base are given up immediately. The belief base account of epistemic states, however, does not *represent* the justificatory structure in the sense that we could read off which of the basic beliefs actually or potentially serve as justifications for which other beliefs. Doyle [47] was striving at such a representation when he devised the first *Truth Maintenance System*.

The representation of the justificatory structure of epistemic states is a desideratum for two different reasons. First, one can argue from a cognitive science perspective that justifications are simply part of the epistemic state of an individual. Justifications are what drives us to accept a meaningful sentence, at least in a wide range of cases. Second, from a computational perspective it appears that revisions can be computed more efficiently when the justificatory structure is available because this structure encodes information as to which derived beliefs are affected in which ways by new epistemic input. Such a structure may also tell us which new beliefs are acquired in response to new epistemic input.

3.3.3 Epistemic Ordering

A sentence α being more entrenched than another sentence β means that α has higher epistemic value in inquiry and deliberation than β. Notably, the ordering of epistemic entrenchment must be determinable independently of future changes of belief sets under new epistemic input [61, p. 87]. This specific requirement could not be met by any explanation of epistemic entrenchment that has been advanced by Gärdenfors and others. The notion of epistemic value of a belief has resisted further clarification and remained too vague to be used in determining entrenchment orderings in non-trivial case studies. Hence, there is no account of measuring epistemic entrenchment orderings (cf. Sect. 3.1.4). The notion of such an ordering, therefore, has remained largely a primitive one in belief revision theory.[4]

Similar considerations apply to the system of spheres by Grove [67] and Spohnian ranking functions [105].[5] This is a problem if an epistemic ordering of beliefs or belief sets is meant to *explain* concrete belief changes of speakers. For this to be seen, it is enlightening to think of an epistemic ordering as introducing some theoretical term into belief revision theory, a view that has been suggested by Rott [98, pp. 43n]:

> The theoretical terms of physics are part of the static picture of physical systems, but by entering into the basic laws of motion, they account for actual and potential changes of such systems. In precisely the same way, orderings of belief bases, for instance, may be regarded as part of the static picture of doxastic systems, but by entering into the basic laws of motion for beliefs, they do account for actual and potential revisions of such systems. This is the reason why *dynamic* are superior to *kinematic* theories.

Let us take this comparison between theoretical terms in physics and belief revision theory a bit further. A theoretical term is a descriptive symbol of a logical language whose meaning is determined through the axioms of one or several

---

[4]With the notable exception of Rott [98, pp. 223–267], who gives an explanation of the relation of epistemic entrenchment in terms of rational choice theory. We will come back to this explanation below.

[5]Hild and Spohn [73] show how ranking functions can be measured, but their proposed method rests on information about how speakers change their beliefs under new epistemic input. This method, therefore, violates Gärdenfors's requirement that the epistemic ordering be determinable independently of dispositions to change beliefs in contractions and revisions.

scientific theories. Among the axioms, Carnap [33] distinguishes between T- and C-postulates: in a T-postulate, every descriptive symbol is a theoretical term, whereas a C-postulate has occurrences of both observational and theoretical terms.

Theoretical terms are essential to predictions in the sciences. This feature strongly depends on theoretical terms having multiple occurrences in the C- and T-postulates. Because of such multiple occurrences, a theoretical term is connected with observational terms in several ways. One of these connections may be used to measure the value of a theoretical term for an empirical object and another for predicting the outcome of an experiment with this object. This is what we do when we measure first the mass of bodies by means of a balance and then predict the behaviour of these bodies in collision experiments. The existence of more than one connection to the observable phenomena is essential to the explanatory power of theoretical terms as well: an expression is of some explanatory value in a certain context if and only if its meaning got partially determined in another context before.

In sum, a proper theoretical term is connected to observable phenomena by means of T- and C-postulates in various ways. With this in mind, we can give our criticism of the conceptual explanations that have been advanced so far for the various orderings of beliefs a more concise formulation: existing conceptions of the epistemic ordering of beliefs are not well connected to observable phenomena other than observations as to what is happening to epistemic states in contractions and revisions. There is at most one C-postulate that relates the ordering of beliefs to phenomena that are observable independently of belief revision theory, i.e., potential belief changes as occurring in revisions and contractions. In the case of epistemic entrenchment, equation (C $\leq$) figures as such a postulate:

$$(\mathrm{C} \leq) \quad \alpha \leq \beta \text{ iff } \alpha \notin K \div (\alpha \wedge \beta) \text{ or } (\alpha \wedge \beta) \in Cn(\emptyset).$$

In addition to such a C-postulate, there are connections between the epistemic ordering and other theoretical concepts. For example, Gärdenfors's original idea to interpret the epistemic entrenchment ordering as an ordering of epistemic value can be seen as establishing a T-postulate between the epistemic entrenchment and the epistemic value of a proposition. Rott [98, Chap. 8] gave an explanation of epistemic entrenchment in terms of rational choice theory and has thus connected epistemic entrenchment with the notion of the good. However, it is only very general axioms that can be shown to hold for belief changes in each variant of an AGM-style belief revision theory, usually some subset of the extended Gärdenfors postulates for revisions and contractions. For example, entrenchment-based revisions can be shown to satisfy K^*1 to K^*8.

In conclusion, we can say that existing conceptions of the epistemic ordering of beliefs allow us to explain very general things about belief changes, but fail to explain the concrete choices that are to be made when epistemic input conflicts with the current belief set or base. In this respect, an epistemic ordering conception is dissimilar to theoretical concepts in the sciences. It seems a desideratum, therefore, to establish more informative connections to observable phenomena for some of these conceptions. This would enrich the explanatory value of the respective epistemic ordering for belief changes. We understand Gärdenfors's requirement that an epistemic entrenchment ordering must be determinable independently of belief changes in contractions and revisions as saying that an epistemic ordering conception must also explain the concrete belief changes that we observe to happen.

Chapter 4
Defeasible Reasoning

This chapter is concerned with defeasible reasoning. Following a brief discussion of the rationale for developing systems of defeasible reasoning, we shall give a detailed account of the preferred subtheories approach by Brewka [23]. This approach will be used to merge structuralism with belief revision theory. Moreover, the truth maintenance system by Doyle [47] deserves consideration since it allows for a representation of the justificatory structure of epistemic states.

4.1 The Rationale

4.1.1 Common-Sense Reasoning

Reasoning with classical logic is *monotonic* in the sense that an extension of the premises of an argument never invalidates any of its conclusions. In more formal terms, monotonicity means that whenever a sentence α is a consequence of a set A of sentences, it is also a consequence of any superset of A:

Definition 4.1 (Monotonicity) A consequence operation $C : \mathcal{P}(L) \rightarrow \mathcal{P}(L)$ is *monotonic* if and only if it holds for all sets A, B of sentences of L: $C(A) \subseteq C(A \cup B)$.

Monotonic logics proved too restrictive to account for common-sense reasoning to the full extent. There, inferential patterns are in use that are only defeasibly valid in the sense that further information may invalidate a conclusion once drawn at a previous state of information. For, commonsensical reasoning relies on defaults, i.e., rules and axioms that are valid only on conditions that most of the time but not always obtain. Take the following well-known examples: if x is a professor, then x teaches; if x is a bird, then x flies. These rules lead to true conclusions quite reliably

H. Andreas, *Dynamic Tractable Reasoning*, Synthese Library 420,
https://doi.org/10.1007/978-3-030-36233-1_4

but do have exceptions. Professors who are on leave do not teach, and penguins are birds but do not fly.

Let us call those conditions under which a certain default rule fails to hold *unfavourable*. Conditions, by contrast, under which the default leads to conclusions that remain accepted are called *favourable* ones. So, being a penguin is an unfavourable condition of the rule that birds fly. Furthermore, we understand the notion of an unfavourable condition to include the negation of the conclusion of a default rule. For example, if Max is a Bavarian, then $\neg Drink\_Beer(Max)$ is an unfavourable condition of the default rule that Bavarians drink beer. Unfavourable conditions are also called *defeaters*, where one distinguishes between directly and indirectly defeating propositions [25, p. 3].

Inferential patterns using defaults are often applied without knowing that unfavourable conditions do not obtain. We may expect a professor to teach without knowing whether or not she is on leave. Hence, people take up an inherent risk of going wrong when using default rules. For this reason, it would be inappropriate to formalise the inference under consideration by means of the following rule:

$$\frac{Professor(x) \wedge \neg Leave(x)}{Teaches(x)}.$$

Hence, the need of systems arises that achieve two things. First, allow one to apply defaults in the absence of premises as to whether favourable or unfavourable conditions obtain. Second, force one to retract the conclusion of a default as soon as an unfavourable condition becomes derivable from the set of accepted sentences.

Such is the essence of defeasible reasoning. Conclusions of inferences may become defeated with an extension of the premises. In more technical terms, defeasible reasoning consists in inferences that do not satisfy the property of monotonicity. As for the terminology, *defeasible* and *nonmonotonic* reasoning are taken as synonyms here.

4.1.2 Scientific Reasoning

Formalising common-sense reasoning is a straightforward motivation for developing and studying nonmonotonic logics. The scope of defeasible reasoning, however, goes beyond everyday contexts. As has been partially observed in the literature [103], defeasible inferences are deeply entrenched in scientific reasoning as well. Two preliminary arguments may be given here why this is so.

With the intent to characterise its nonmonotonic character, Rott [98, p. 32] describes everyday reasoning as ampliative, adventurous and risky. A nonmonotonic inference is ampliative because the content of the conclusion is not contained in the content of the premises. There are worlds (classical interpretations of the respective formal language) in which the premises are true, while the conclusion is false. Such an inference is adventurous and risky since, by transcending the content of the premises, it may lead us to conclusions that need to be retracted at a later stage of inquiry. If so, the adventure of making the inference has failed. The conclusion, however, may also prove both consistent and valuable, in which case the adventure of making the inference would have succeeded.

Is scientific reasoning purely deductive or, at least partly, adventurous and risky? We should note that scientific reasoning consists, to a large extent, of transitions from empirical phenomena to theoretical descriptions thereof. The theoretical description commonly has a richer content than the empirical one in that it entails, in the context of some background theory, sentences that are not entailed by the empirical description, where entailment is understood in the sense of classical logic. For example, assertions about the mass of an object entail, unlike assertions about the object's space-time function, sentences about the object's behaviour in potential future experiments, such as collisions with other objects. The theoretical description, therefore, excludes worlds (classical interpretations) that are not excluded by the empirical description. In this sense, scientific reasoning is ampliative and hence also adventurous and risky. The theoretical description may become retracted if it does not fit well into related empirical phenomena and theoretical interpretations thereof. It will be spelled out in detail, by means of the structuralist framework, what it is for a theoretical description of an empirical phenomenon to fit into the theoretical accounts of related empirical phenomena.

There is another argument for the presence of defeasible inferences in science, which brings the notion of an explanation to bear. Through the transition from an empirical description of a phenomenon to a theoretical description, we obtain an explanation of that phenomenon. Now, advancing an explanation is often a competitive affair in that we must evaluate which of the of rival explanations is superior to the rest. Superiority among different explanations strongly depends on the state of the art of background theories and on further empirical findings. Since background theories may change, novel theories may be advanced and since the set of empirical findings is continually growing, novel explanations can become available that supersede older ones. In this case, a given explanation is defeated by a new one. It seems promising, therefore, to account for the competition of available explanations by means of nonmonotonic reasoning.

4.1.3 Consequence vs. Inference Operations

There is an influential proposal on how to indicate whether the logic under consideration is monotonic or not: the notion of a *consequence* operation is to be reserved for monotonic systems, whereas the corresponding notion of an *inference* operation applies to nonmonotonic systems [98, pp. 31–34]. Wherever possible, this convention has been followed here. To express that a sentence α is a logical consequence of a set A of sentences and that α can be inferred from A, we will write

$$A \models \alpha \quad \text{and} \quad A \mathrel{|\!\sim} \alpha$$

respectively. This is the common relational notation of consequences and inferences, which complements the operational notations $Cn(A)$ and $Inf(A)$. It holds by definition that

$$A \models \alpha \text{ iff } \alpha \in Cn(A) \quad \text{and} \quad A \mid\!\sim \alpha \text{ iff } \alpha \in Inf(A).$$

4.2 Preferred Subtheories

4.2.1 Basic Ideas

The construction of preferred subtheories by Gerhard Brewka [23, 24] is a highly intuitive means to formalise defeasible reasoning. We will refer to this formalism as the *preferred subtheories approach* to defeasible reasoning, also abbreviated by *PSA*. It is guided by the following two ideas:

(1) We can derive a consistent set of conclusions from a potentially inconsistent set S of premises by constructing *maximal consistent subsets* of S.
(2) Sets of premises are ranked. In case of a logical conflict between two premises from different epistemic ranks, the premise of the higher rank is preferred over the other. That is, only the premise at the higher rank can be used for drawing inferences.

The notion of a maximal consistent subset of a given set S is defined as follows:

Definition 4.2 (Maximal consistent subset) Let S be a set of sentences. S' is a maximal consistent subset of S if and only if

(1) S' is classically consistent
(2) $S' \subseteq S$
(3) there is no set S'' such that (i) S'' is classically consistent and (ii) $S' \subset S'' \subseteq S$.

In less formal terms, a maximal consistent subset of a set S is a consistent subset of S that cannot be extended by adding further sentences from S without becoming inconsistent.

4.2.2 Inferences from Preferred Subtheories

Let us move on to defining the inference relation of the preferred subtheories approach to defeasible reasoning[1]:

Definition 4.3 (Default theory) A default theory is a sequence $\langle T_1, \ldots, T_n \rangle$ such that any member of any set T_i $(1 \leq i \leq n)$ is a first-order sentence.

[1]The present exposition is based on Brewka [23] and Brewka et al. [25].

The different components of such a sequence are intended to represent different degrees of *reliability*. Beliefs of T_1 are considered to have the highest degree, whereas beliefs of T_n are assigned the least degree. If there are any hard facts and conceptual truths, these are members of the set T_1. Here is a graphical way of conceiving the modular ordering of beliefs as encoded by a default theory:

T_1
T_2
...
T_n

Universally quantified and open formulas are to be replaced by their ground instances in any set T_i $(1 \leq i \leq n)$. So $F(x) \rightarrow G(x)$ is to be replaced by $F(t_1) \rightarrow G(t_1)$, $F(t_2) \rightarrow G(t_2)$, ...; where $t_1, t_2, \ldots$ are closed individual terms. For the sake of simplicity, it is still customary to write $F(x) \rightarrow G(x)$, but this formula is to be read as a representation of its ground instances. This style of representing a universal formula allows one to make selective use of its instances. It is this option of selectively using instances of a generalisation by means of which Brewka [24] refined the framework of Rescher [95].

The union of $T_1, \ldots, T_n$ need not be consistent. In most interesting applications, it fails to be so. Therefore, inconsistencies are to be removed by the inference formalism. For this to be achieved, *maximal consistent subsets* of $T_1 \cup \ldots \cup T_n$ are to be chosen. This choice is constrained by the levels of reliability so as to ensure that, in case of a conflict between sentences with different degrees of reliability, the more reliable ones are preferred. The notion of a preferred subtheory makes this idea precise:

Definition 4.4 (Preferred subtheory) A set $S = S_1 \cup \ldots \cup S_n$ is a preferred subtheory of a default theory $\Theta = \langle T_1, \ldots, T_n \rangle$ if and only if, for all i $(1 \leq i \leq n)$, $S_1 \cup \ldots \cup S_i$ is a maximal consistent subset of $T_1 \cup \ldots \cup T_i$.

So the construction of a preferred subtheory starts with a maximal consistent subset of T_1, goes on to add as many formulas of T_2 as can consistently be added, and so on. This construction rarely results in a uniquely determined preferred subtheory. Rather, there are usually several sets of formulas that qualify as a preferred subtheory of a default theory Θ. A standard way to establish an inference relation upon such a set of preferred subtheories is to take just those formulas as inferable which are logical consequences of any preferred subtheory:

Definition 4.5 (Sceptical inference relation in PSA) Let Θ be a default theory and P the set of preferred subtheories of Θ. ϕ is an inference of Θ – in symbols: $\Theta \mid\!\sim^{PS(s)} \phi$ – if and only if $\phi \in \bigcap_{S \in P} Cn(S)$.

Cn denotes the operation of logical consequence in classical logic, as is standard. The formalism of preferred subtheories qualifies as a logic insofar as it defines an inference relation for sets of formulas of a formal language with the standard syntax of first-order logic. A fully fledged proof theory has been developed for related systems in the framework of adaptive logics (see Meheus et al. [83]).

Sceptical reasoning means that the inferential closure of Θ is defined by the intersection of the sets $Cn(S)$. This contrasts with the *credulous* inference relation, the inferential closure of which is defined by the union of the sets $Cn(S)$. (S ranges over the preferred subtheories of Θ.) The inferential closure may well be inconsistent when we adopt the credulous definition. This is why credulous inferences in PSA are not suited to define belief changes.

The just defined sceptical inference relation can be made even more sceptical. Instead of defining that ϕ is inferable from a default theory Θ if and only if it is a classical consequence of all preferred subtheories of Θ, one may put forward the somewhat stronger definition, according to which ϕ must be a classical logical consequence of the intersection of all preferred subtheories (cf. [83]):

Definition 4.6 (Super-sceptical inference relation in PSA) Let Θ be a default theory and P the set of preferred subtheories of Θ. ϕ is a super-sceptical inference of Θ – in symbols: $\Theta \mid\!\sim^{PS(su)} \phi$ – if and only if $\phi \in Cn \bigcap_{S \in P} S$.

This definition simply varies the order in which the operations of intersection and logical consequence are performed upon preferred subtheories. Compared to sceptical reasoning, super-sceptical reasoning has the benefit of avoiding the acceptance of sentences that are based on contradictory arguments. For example, ϕ may be a consequence of both S_1 and S_2 but not of $S_1 \cap S_2$, where S_1 and S_2 are the only preferred subtheories. In this case, S_1 and S_2 contradict one another since, otherwise, they would not be different preferred subtheories. In view of this contradiction, one should carefully assess whether ϕ deserves in fact acceptance.

4.2.3 A Simple Example

It is illuminating to see how a standard example of default reasoning works with preferred subtheories. Clearly, most birds are capable of flying. Hence, it is reasonable – though not classically valid – to infer from a being a bird that a can fly. We have to retract this inference, however, once we come to know that a is a penguin. Now consider the following default theory Θ:

$T_1 : Bird(a)$
$T_2 : Penguin(a) \rightarrow \neg Flies(a)$
$T_3 : Bird(a) \rightarrow Flies(a)$

Θ has exactly one preferred subtheory $S = \{Bird(a), Penguin(a) \rightarrow \neg Flies(a), Bird(a) \rightarrow Flies(a)\}$. Hence, it holds that $\Theta \mid\!\sim Flies(a)$. Suppose we come to know, from a highly reliable source, that a is a penguin. The new epistemic state is represented by a default theory Θ':

T_1 :	$Bird(a), Penguin(a)$
T_2 :	$Penguin(a) \rightarrow \neg Flies(a)$
T_3 :	$Bird(a) \rightarrow Flies(a)$

Θ' has exactly one preferred subtheory $S = \{Bird(a), Penguin(a), Penguin(a) \rightarrow \neg Flies(a)\}$. Hence, $\Theta' \not\mid\!\sim Flies(a)$; which means that the inference to $Flies(a)$ is retracted. This behaviour of the inference relation $\mid\!\sim$ is also described as *nonmonotonicity*.

4.2.4 Belief Revision with Preferred Subtheories

Recall the basic ideas of the direct mode of belief base revision, explained in Sect. 3.2.3: revisions are made in a plain and artless fashion, simply by adding the new epistemic input to the belief base. The formation of the belief set $K(H)$ of a belief base H is governed by a nonmonotonic and paraconsistent inference operation. The notion of the direct mode of belief base revision in Rott [98, Chap. 1] generalises the definition of belief changes in the framework of preferred subtheories in Brewka [24]:

Definition 4.7 **($K(\Theta)$ and $K(\Theta) *_i \alpha$)** Let Θ be a default theory. The formation of the belief set $K(\Theta)$ and its revision are defined as follows:

$$Inf(\Theta) = \{\phi \mid \Theta \mid\!\sim^{PS} \phi\}$$

$$K(\Theta) = Inf(\Theta)$$

$$K(\Theta) *_i \alpha = Inf(\Theta \oplus_i \alpha)$$

where $\oplus_i \alpha$ designates the operation of adding α to Θ at level i.

Note that the epistemic input consists of a revision formula and the epistemic rank of this formula, where this rank is relative to the other elements in Θ. Hence, Brewka [24] defines a revision operation $*_i$. $\mid\!\sim^{PS}$ is placeholder that stands for the sceptical or the super-sceptical inference relation.[2]

Revisions in PSA have some noteworthy properties, which are easy to prove [24]:

Proposition 4.1 *Revisions in PSA do not always satisfy the success postulate K^*2.*

Proposition 4.2 *Revisions in PSA violate K^*5.*

Proposition 4.3 *Revisions in PSA are commutative, i.e., for all sentences α, $\beta \in L$: $(K * \alpha) * \beta = (K * \beta) * \alpha$.*

[2] We leave out the definition of contractions because they are not needed as an auxiliary notion in the definition of revisions.

Violation of the success postulate can be reasonable in case both the source of the new information α is not reliable and α contradicts other well established beliefs. Think of an article in the yellow press reporting there has been no global warming in the past 100 years. Think of an experiment whose outcome contradicts very well established universal laws. Success is particularly counter-intuitive if our epistemic input consists of a logical falsehood.

Universal satisfaction of success makes sense when the epistemic input is confined to the phenomenal core of the new information, quite in the spirit of old-fashioned phenomenalism. This would mean that new epistemic input had to have the form of reports like: I read in the yellow press that ϕ; my visual senses delivered to me that ϕ; I measured by means of such and such an experimental equipment that ϕ. Is such a constraint upon the content of the epistemic input a viable option in the AGM theory? The answer depends on whether K may contain such sentences as 'when I read ϕ in the newspaper, then ϕ', 'when my senses deliver ϕ to be the case, then ϕ', etc. If such axioms are given as universally quantified formulas, they will be refuted soon and thus would not remain in K. If, by contrast, we choose to represent axioms of the background theory by all of their ground instances, we can selectively apply such axioms. With this representation of axioms, confining the epistemic input to the phenomenal core of new information becomes a viable option in the AGM theory.

(K*5) says that the absurd belief state, which contains all formulas of L, is entered if and only if the epistemic input consists of a logical falsehood. Such an epistemic input is dealt with somewhat differently in PSA. Even though a logical falsehood may well be adopted at some level in Θ, no preferred subtheory will ever contain a sentence that is false in virtue of logic. The absurd belief state $K_{\perp}$, therefore, is avoided altogether. It seems almost conceptually true that the absurd belief state better be avoided if this can be achieved without particularly high costs.

Commutativity of the epistemic input is highly intuitive when studying belief changes in scientific contexts. Assume there are two experiments e_1 and e_2 such that the result of the first is given by proposition α_1 and that of the latter by α_2. α_1 and α_2 may contradict one another. Further, assume e_1 and e_2 are considered equally reliable. In this scenario, acceptance of α_1 and α_2 ought not to depend on the order in which e_1 and e_2 are conducted and interpreted. To give a simple example, consider measurement of a the pH-value of a liquid a. Assume the measured values of two experiments strongly diverge from one another, say we measured first 4,5 and then 8,9. Then, it is not rational at all to accept the proposition that pH-value(a)=8,9. Nor is it reasonable to accept the conjunction pH-value(a)=8,9 $\wedge$ pH-value(a)=4,5 because this would force us to retract the belief that there is a quantity called *pH-value* that has the formal properties of a function. We conclude, therefore, that either the success postulate must be given up or the epistemic input be confined to its phenomenal core. In the present example, this would amount to accepting that it was measured that pH-value(a)=8,9 in place of accepting pH-value(a)=8,9.

There remains, of course, the problem that the distinction between the phenomenal core of a statement like 'I read that α', and the statement itself is a relative one. Arguably, the statement that I read that α has its own phenomenal core, which may be captured by the statement that my visual sense delivered to me that $\ulcorner\beta\urcorner$ is written in the newspaper and that I understand $\ulcorner\beta\urcorner$ to mean α. This relativity of the presumed phenomenal core of a statement or hypothesis is taken into account in the structuralist approach, as we shall come to observe.

4.2.5 Reversing the Levi Identity

The belief revision scheme of preferred subtheories is not quite as well received as other approaches to belief base revision, such as partial meet base revision. Some handbook articles and the textbook by Hansson [69] do not consider it. However, belief changes defined by preferred subtheories fit quite nicely into the framework of Hansson's "Reversing the Levi Identity" [68], which is foundational to the study of belief bases in the AGM tradition. There, Hansson distinguishes between internal and external revisions of a belief base. In formal terms:

$$H * \alpha = (H \div \neg\alpha) \cup \{\alpha\}$$

$$H \pm \alpha = (H \cup \{\alpha\}) \div \neg\alpha.$$

The former operation is called *internal* since the contraction by the negation of the revision formula is performed inside the original belief base [68, p. 649]. The latter revision operation is called *external* since the contraction is performed for the extended belief base, extended by the revision formula α.

Suppose we understand the expansion of a belief base H as a simple set-theoretic addition:

$$H + \alpha = H \cup \{\alpha\}.$$

Then, we can say that internal base revisions follow the Levi identity (Def $*$). External base revisions, by contrast, *reverse* the order of operations of the Levi identity: we first expand the belief base, then it is contracted.

Now, suppose the epistemic rank i of the revision formula α – in a revision $K(\Theta) *_i \alpha$ – is such that $T_1 \cup \ldots \cup T_i \not\models \neg\alpha$. Then, belief changes defined by the super-sceptical inference relation of preferred subtheories can be obtained by the following sequence of operations: (i) expand the ranked belief base by α, (ii) contract the ranked belief base by $\neg\alpha$, and then (iii) close the contracted belief base by the operation of logical consequence in classical logic. Hence, if the revision by α is successful – in the sense that $\alpha \in K(\Theta *_i \alpha)$ –, then $K(\Theta *_i \alpha)$ equals the logical closure of an external belief base revision.

4.3 Truth Maintenance

The truth maintenance system by Doyle [47] deserves noticing for several reasons. First, it qualifies as a frame system in the sense of Minsky [87]. Second, it is non-monotonic but does not require computing derived beliefs from scratch whenever new input comes up. Third, it keeps track of justifications for beliefs. Fourth, it is guided by a broadly foundationalist epistemology, as has been emphasised by Doyle himself [49]. We consider all four of these features highly desirable and cognitively plausible, and will endeavour to realise them in the system to be devised in the final chapters of this investigation.

The original truth maintenance paper in [48] is very rich in content, ranging from Minsky's criticism of the logistic approach to an attempt at a formalisation of dialectics, which is conceived as a method of argumentation in philosophy. In the present exposition, we confine ourselves to the core of Doyle's system, following [49] and [25, Chap. 4]. A truth maintenance system (TMS henceforth) consists of a set of nodes N and a set of justifications J. Nodes can take on two states, IN and OUT. The IN state corresponds to acceptance or truth, the OUT state to rejection or falsehood. A node represents a proposition whose inner logical structure is not represented. A labelling L is a mapping from the set of nodes to the set of states. Justifications have the following notation:

$$\langle \alpha_1, \ldots, \alpha_n \mid \beta_1, \ldots, \beta_m \rightarrow \gamma \rangle.$$

The meaning of such a justification is: if $\alpha_1, \ldots, \alpha_n$ are IN and $\beta_1, \ldots, \beta_m$ OUT, then γ should be IN. So, $\alpha_1, \ldots, \alpha_n, \beta_1, \ldots, \beta_m$ and γ designate nodes. $\alpha_1, \ldots, \alpha_n$ is called the *inlist*, $\beta_1, \ldots, \beta_m$ the *outlist* and γ the consequent.

A labelling of the nodes must satisfy two conditions in order to be admissible. First, for all justifications $j \in J$, if the nodes of the inlist are IN and those of the outlist OUT, the consequent must be IN (closedness). Second, for all nodes being IN, there must be a non-circular justification by means of members of J (groundedness). Of course, for there to be grounded nodes, there must be premises that have no further justification. Such premises can be specified by justifications with an empty inlist. These ideas can be made precise as follows [25, Chap. 4]:

Definition 4.8 (Valid justification) Let N be a set of nodes in a set of justifications J and $L : N \rightarrow \{IN, OUT\}$ a labelling. A justification

$$\langle \alpha_1, \ldots, \alpha_n \mid \beta_1, \ldots, \beta_m \rightarrow \gamma \rangle$$

is valid for γ in L if and only if, for all α_i $(1 \leq i \leq n)$, $L(\alpha_i) = IN$ and for all β_j $(1 \leq j \leq m)$, $L(\beta_j) = OUT$.

Definition 4.9 (Closedness) A labelling L is closed if and only if $L(\gamma) = IN$ whenever J contains a justification for γ that is valid in L.

Definition 4.10 (Groundedness) A labelling L is grounded if and only if there is a linear ordering $\langle \phi_1, \phi_2, \ldots \rangle$ of the set

$$INS(L) = \{\gamma \in N \mid L(\gamma) = IN\}$$

such that every node $\phi_i \in INS(L)$ has a valid justification

$$\langle \delta_1, \ldots, \delta_n \mid \epsilon_1, \ldots, \epsilon_m \rightarrow \phi_i \rangle$$

in L such that $\{\delta_1, \ldots, \delta_n\} \subseteq \{\phi_1, \ldots, \phi_{i-1}\}$.

Definition 4.11 (Admissibility of a labelling) A labelling L is admissible if and only if it is both closed and grounded.

There is usually more than one admissible labelling because applying justifications in a different order may result in another labelling. Hence, there is a credulous and a sceptical approach to reasoning. If we are credulous, a node is believed if and only if there is some admissible labelling in which it is IN. If, by contrast, we are sceptical, a node is believed if and only if it is IN on all admissible labelings. In the literature it is sometimes assumed that we need to compute all admissible labelings, which is in fact necessary for some applications.[3]

Notably, Doyle himself did not feel the need to compute all admissible labelings. His view was rather that a certain determinate and admissible labelling represents (together with the set of justifications) *the* epistemic state of an individual. Revisions of such states, then, consist in adjusting the labelling in the face of new epistemic input, where the input always has the form of a new justification. Doyle developed a sophisticated algorithm for the transition of admissible labelings under new epistemic input. This algorithm is at the core of the original TMS as published in [47]. The approach by Doyle is thus credulous, with the qualification that only one admissible labelling is actually computed. Strictly speaking, the approach is neither credulous nor sceptical.

A TMS qualifies as a frame system in the sense of Minsky: a justification has slots with admissible values (IN and OUT) and a complex condition that the values of these slots must satisfy (if the inlist nodes are IN and the outlist nodes OUT, then the consequent must be IN). Notice, further, that the respective inlists, outlists and conclusions of justifications usually overlap. In this sense, the justifications of a TMS form a frame system.

The process of giving justifications is nonmonotonic because a node of the outlist of a valid justification may change from OUT to IN at a later stage of reasoning, in which case this justification would be invalidated. Unlike in other nonmonotonic systems, derived beliefs need not be computed from scratch since the TMS algorithm tries to adjust, in the face of new epistemic input, previous

---

[3]Because of the obvious correspondence between defaults in default logic and justifications in Doyle's TMS, the latter has been used for computing extensions in default logic [25, Chap. 4.4]

"derivations" such that the new incoming reason is taken into account. Previously derived beliefs have the form of nodes being labelled IN in the present state of the TMS.

The justificatory structure of the epistemic state is represented in the TMS insofar as all reasons in play are recorded. There is, however, not necessarily bookkeeping of how we came to hold certain beliefs. That the TMS is guided by foundationalist principles should be obvious.

As indicated in the Introduction, we shall generalise Doyle's original TMS in three respects. First, our TMS will be more liberal concerning the logical form of what Doyle calls *justifications*. Any instance of an axiom can be a justification. Second, justifications may well become retracted. Third, there is an epistemic ranking of justifications. Unlike the original TMS by Doyle, our TMS does not accept mere credulous inferences. Its inference relation is rather super-sceptical, as we shall see.

Part II
Belief Revision with Frames

Chapter 5
Postulates for Structuralism

In this chapter, we develop a novel account of structuralist theory representation. Its novelty derives from using Carnapian postulates and a refined semantics of theoretical terms. In essence, we devise a system of postulates for a theory-net in structuralism. This system of postulates can be seen as an axiomatic account of reasoning with frame concepts. The account is static, but will be merged with belief revision theory in the next chapter.

5.1 Preliminaries

Recall from Chap. 2.3 that a theory-element is the simplest unit of theory representation in structuralism. Three sets of structures are always constitutive of a theory-element $\mathbf{T}$:

(1) $\mathbf{M_p}(\mathbf{T})$, the set of potential models of $\mathbf{T}$
(2) $\mathbf{M}(\mathbf{T})$, the set of actual models of $\mathbf{T}$
(3) $\mathbf{I}(\mathbf{T})$, the set of intended applications of $\mathbf{T}$.

Models of $\mathbf{T}$ are those systems that satisfy the substantial law of $\mathbf{T}$, whereas potential models only satisfy certain frame conditions under which it makes sense at all to ask whether the substantial law of $\mathbf{T}$ holds for a structure x. The law of conservation of momentum, for example, can only be applied to structures that contain relations of the types of the velocity and the mass function.

$\mathbf{M}(\mathbf{T})$, $\mathbf{M_p}(\mathbf{T})$ and $\mathbf{I}(\mathbf{T})$ are symbols for extensions of set-theoretic predicates. The choice between (set-theoretic) predicate symbols and symbols for extensions of such predicates, however, is inessential. For example, we could equally well write $\mathbf{M}(\mathbf{T})(\mathbf{x})$ in place of $x \in \mathbf{M}(\mathbf{T})$. In the former notation $\mathbf{M}(\mathbf{T})$ designates a predicate, while the latter designates an extension of a predicate. Definitions of extensions of set-theoretic predicates thus also define corresponding predicates.

H. Andreas, *Dynamic Tractable Reasoning*, Synthese Library 420,
https://doi.org/10.1007/978-3-030-36233-1_5

Formal definitions of $\mathbf{M(T)}$ and $\mathbf{M_p(T)}$ are essential to a structuralist representation of a theory-element $\mathbf{T}$. A definition of $\mathbf{M(T)}$ differs from one of $\mathbf{M_p(T)}$ in that it contains at least one law. That is, if P is a set-theoretic predicate expressed by formulas $\phi_1, \ldots, \phi_s$, then there is at least one law among these formulas. If, by contrast, P defines the potential models of $\mathbf{T}$, then there are no laws among the formulas $\phi_1, \ldots, \phi_t$ that define P. The conjunction of laws in the definition of $\mathbf{M(T)}$ is called the *substantial law* of $\mathbf{T}$. For many theory-elements $\mathbf{T}$, there is a single axiom which is the substantial law of $\mathbf{T}$.

What is a theory-element itself? We do not attempt at a formal definition of this notion. It is merely important to note that a theory-element allows us to refer to a substantial law. Furthermore, we assume that any substantial law closely corresponds to a law, or a conjunction of laws, of a scientific theory – in the intuitive sense of this term.

Matters of syntax remain more or less implicit in a (classical) structuralist representation of scientific theories. The formulation of laws and characterisations in a definition of models requires logical, set-theoretic, mathematical and descriptive vocabulary. It is assumed that mathematical concepts are reducible to logical and set-theoretic ones, but the representation of scientific theories remains semiformal in the sense that mathematical concepts are used directly without reduction to logic and set theory. Hence, we distinguish between logical, set-theoretic, mathematical and descriptive symbols, and concepts.

In the course of this investigation, it will occasionally be assumed that the syntax of the language of a theory-element $\mathbf{T}$ has been made explicit. The set of well-formed formulas of this language will be referred to by $L(\mathbf{T})$. The rules for the formation of formulas will be that of first-order logic. $L(\mathbf{T})$ must be such that all conditions in the definitions of $\mathbf{M(T)}$ and $\mathbf{M_p(T)}$ are expressible. So, for example, if formulation of the substantial law of $\mathbf{T}$ requires certain mathematical symbols, these must be part of the mathematical vocabulary of $L(\mathbf{T})$.

As has been pointed out by Suppes [110, Chap. 2], the semiformal style of defining set-theoretic predicates is precise enough to indicate how a complete formalisation using axiomatic set theory can be obtained. The axiomatisation of set theory in first-order logic is assumed to satisfy the following conditions:

(1) There is one and only one domain of interpretation, which comprises empirical objects and sets.
(2) The axiomatisation of set theory must make room for *Urelements* in order to distinguish empirical objects from one another.

Several systems satisfying these two conditions have been developed in the literature. For an extension of ZF that accounts for distinguishable Urelements see Löwe [81]. Another promising strategy is to use the formal axiomatisation of naive set theory in a paraconsistent framework, such as that of adaptive logics (see Verdée [114]).

5.2 Intended Applications and Theoreticity

Theoreticity of concepts is a key theme in structuralism. A number of different and quite sophisticated criteria of how to draw the distinction between theoretical and non-theoretical concepts have been suggested since the seminal work of Sneed [104] appeared. All of these are driven by two ideas. First, the notion of theoreticity is relativised to a theory-element so that we always have to distinguish between **T**-theoretical and **T**-non-theoretical concepts as opposed to using the unqualified distinction between theoretical and observational concepts. Second, a concept C is **T**-theoretical only if some axioms of **T** are indispensable for determining the extension of C. The division into **T**-non-theoretical and **T**-theoretical concepts leads to a corresponding distinction between **T**-theoretical and **T**-non-theoretical structures:

$$\langle D_1, \ldots, D_k, A_1, \ldots, A_m, N_1, \ldots, N_p \rangle$$

$$\langle D_1, \ldots, D_k, A_1, \ldots, A_m, N_1, \ldots, N_p, T_1, \ldots, T_q \rangle.$$

$T_1, \ldots, T_q$ designate extensions of **T**-theoretical concepts, whereas $N_1, \ldots, N_p$ stand for extensions of **T**-non-theoretical concepts. In the case of classical mechanics we have:

$$\langle P, T, S, \mathbb{N}, \mathbb{R}, c_1, c_2, s \rangle \tag{5.1}$$

$$\langle P, T, S, \mathbb{N}, \mathbb{R}, c_1, c_2, s, m, f \rangle. \tag{5.2}$$

Intended interpretations of the symbols used to designate these structures are as follows. P is a set of particles, T a temporal interval, S a set of space points, c_1 a function assigning real numbers to time points, c_2 a function assigning real numbers to space points. s is the space function, m the mass function and f the force function.

An intended application is a **T**-non-theoretical structure in the sense that it comprises only sets representing the extension of T-non-theoretical concepts, in addition to the base sets $D_1, \ldots, D_k$, and $A_1, \ldots, A_m$. Thus, in the case of classical mechanics, an intended application is a structure of type (5.1). Intended applications are the phenomena to which a theory is applied, where the theory-phenomenon distinction is always relativised to a theory-net or theory-element.

For a **T**-non-theoretical structure to be an intended application, it must be extensible to a **T**-theoretical structure that is a potential model of **T**. To make this requirement precise, we need to introduce a restriction function **r**(**T**) that "cuts off" the **T**-theoretical concepts from a **T**-theoretical structure to obtain a **T**-non-theoretical structure:

Definition 5.1 (**r**(**T**)) Let x be a structure of type $\langle D_1, \ldots, D_k, A_1, \ldots, A_m, N_1, \ldots, N_p, T_1, \ldots, T_q \rangle$. $y = \mathbf{r}(\mathbf{T})(x)$ if and only if (i) y is a structure type $\langle D_1, \ldots, D_k, A_1, \ldots, A_m, N_1, \ldots, N_p \rangle$ and (ii) for all i $(1 \leq i \leq k + m + p)$ $(x)_i = (y)_i$.

$(x)_i$ always designates the i-th component of a structure x. Using the restriction function $\mathbf{r}(\mathbf{T})$, we can give a formal characterisation of the empirical systems that are intended applications of a theory-element **T**:

Explanation 5.1 If $y \in \mathbf{I}(\mathbf{T})$, then there is an x such that $x \in \mathbf{M_p}(\mathbf{T})$ and $y = \mathbf{r}(\mathbf{T})(x)$.

In other words, an intended application must be obtainable from a potential model x by restricting x to a **T**-non-theoretical structure. This requirement, however, is only a necessary condition for a structure y being an intended application. It is an assumption inherent in structuralism that the set **I(T)** cannot be characterised completely by formal means. Rather the notion of an intended application is to be taken in the literal sense that there are scientists who think that a theory is applicable to certain empirical systems. Among other things, paradigmatic examples play an important role in determining the range of such empirical systems. Henceforth, we assume that the extension of ⌜**I(T)**⌝ is given somehow. The information encoded by the set **I(T)** is comparable to a complete interpretation of the observation language in non-set-theoretic accounts of scientific theories.

5.3 Intended Applications Liberalised

It is beyond doubt that with the advancement of a relative criterion of theoreticity important progress has been made in the analysis of theoretical concepts. Less obvious, however, is whether an intended application of **T**, in a more intuitive and cognitive sense of the term, must always have the precise type of a **T**-non-theoretical structure. Arguably, there are cases where the **T**-non-theoretical concepts of an empirical system to which **T** is applied are not determined independently of **T**. Take for example measurement of time. For a long time, the rotation of the Earth served as a standard of measurement. This standard was underlying the proposition that the angular velocity of the rotation of the Earth is constant, which can be derived from the law that the angular momentum of a rotating body remains constant in the absence of any outer forces. Since this law is the equivalent of Newton's first law of motion for rotational motions, it should be considered as part of classical mechanics. Hence, we must say that a theory-element of classical mechanics was in fact used to determine the extension of a **T**-non-theoretical concept, viz., the extension of the concept of time. This claim holds in an even more obvious way for the measurement of time by means of a pendulum clock.

Another type of cases where intended applications of a theory-element **T** do not seem to have the precise type of a **T**-non-theoretical structure are predictions. In Carnapian and other accounts of scientific theories predictions are thought of as derivations of observation sentences from theoretical antecedent conditions with the help of axioms of a theory T. In structuralism, we can think of such predictions as consisting in the determination of **T**-non-theoretical concepts using theory-elements of a theory-net **N**. For example, knowing the weight of two bodies a_1 and a_2 lets us predict, with the help of **LP**, for which values of the distance function $n(a_1)$ and $n(a_2)$ are in an equilibrium when placed on a lever. This is a clear case where an intended application of **T** is not a **T**-non-theoretical structure. On the other hand, there are empirical systems to which **LP** is applied that are in fact used to determine the weight of bodies from observed states of equilibrium and not vice versa. The problem surfacing is that the structure type of intended applications of a theory-element **T** exhibits a certain variety.

Arguably, there is a certain tension between the cognitive dimension of the notion of an intended application and its formal explication in classical structuralism. An intended application of **T**, in the cognitive sense, is an empirical system to which the substantial law of **T** is either actually applied in calculations or claimed to hold. Intended applications in this sense show a variety of structure types for a single theory-element, contrary to the formal explication of an intended application in classical structuralism.

For obvious reasons, we are particularly interested in the cognitive dimension of the notion of an intended application. This gives rise to a more liberalised understanding of this notion, according to which an intended application is a (specific type of) substructure of a potential model of **T**. That is, a structure y that is an intended application of **T** must be extensible to a potential model of **T**. This was also required in classical structuralism but with a more rigid notion of an extension. Let us therefore define the notion of a **T**-extension in a more liberal way:

Definition 5.2 (T-extension) Let **T** be a theory-element whose potential models are of the type $\langle D_1, \ldots, D_k,\ A_1, \ldots, A_m,\ R_1, \ldots, R_n \rangle$. Let y be a sequence of this type as well. $EXT(\mathbf{T})(y) = \{x \mid$ (i) for all i $(1 \leq i \leq k+m)$ $(y)_i = (x)_i$, (ii) for all i $(k+m < i \leq k+m+n)$ $(y)_i \subseteq (x)_i$ and (iii) $x \in \mathbf{M_p}(\mathbf{T})\}$. We say that x is a **T**-extension of y if and only if $x \in EXT(\mathbf{T})(y)$.

Defining **T**-extensions of a structure y in this manner makes room for a greater variety of structure types among the intended applications of a theory-element **T**:

Explanation 5.2 If $y \in I(\mathbf{T})$, then $\mathrm{EXT}(\mathbf{T})(y) \neq \emptyset$.

Again, this proposition spells out only a necessary condition for a structure y to be an intended application of **T**. For this to be the case, y must represent an empirical system to which **T** is in fact applied or considered applicable. As noted above, this further requirement cannot be formalised.

The formal characterisation of an intended application by Explanation 5.2 contains two requirements. Concerning the values of the slots for base sets, it holds that an intended application must agree with any of its **T**-extensions. Concerning the values of the relation concepts $R_1, \ldots, R_n$, it must hold that the valuation by an intended application b is a subset of the corresponding valuation by **T**-extensions of b. Loosely speaking, intended applications of **T** are substructures of potential models of **T**.[1]

[1]This understanding of a substructure, even though it is related to, does not fully conform to the understanding of a substructure in model theory. There, substructures are commonly understood as

It should be obvious by now in what ways the present notion of an intended application is more versatile than the original one, whose formal characterisation is given by Explanation 5.1. An intended application in the sense of Explanation 5.2 specifies an interpretation for all relation slots of **T**, where this interpretation may well be a partial one for each slot. On the original understanding, by contrast, an intended application specifies a complete interpretation of the **T**-non-theoretical terms, while the **T**-theoretical terms remained uninterpreted. Still, intended applications that have in fact the form of a **T**-non-theoretical structure (on the original understanding of theoreticity) can be translated to the present conception as follows:

$$\langle D_1, \ldots, D_k, A_1, \ldots, A_m, N_1, \ldots, N_p, \emptyset, \ldots, \emptyset\rangle$$

where the number of empty sets by which an intended application $\langle D_1, \ldots, D_k, A_1, \ldots, A_m, N_1, \ldots, N_p\rangle$ is extended must equal the number of **T**-theoretical concepts, of course.

Even if a relation symbol R_i is **T**-non-theoretical (on the original understanding of theoreticity), its value in an intended application y may be the empty set. Conversely, if R_i is **T**-theoretical (on the original understanding), its value in an intended application of **T** may be a complete function. As a limiting case, an intended application may be just a potential model of **T**. This makes sense since sometimes we have complete information about the properties of an empirical system and think, at the same time, that the substantial law of some theory-element holds true for this system.

5.4 The Postulates

Applying a scientific theory to an empirical system allows us to make novel assertions about the properties of this system. These novel assertions serve the purpose of explaining phenomena and making predictions. Structuralism has it that assertions resulting from the application of a scientific theory are encoded by theoretical extensions of intended applications. To be more precise, applying a scientific theory **T** to an empirical system y aims at a set $\mathbf{A(T)}(y)$ of **T**-extensions of y such that the following properties are satisfied. First, for all $x \in \mathbf{A(T)}(y)$, x is a model of **T**. Second, all $x \in \mathbf{A(T)}(y)$ must satisfy all links (internal, external and specialisation links) to $\mathbf{T}'$-extensions of other intended applications, where **T** is identical to $\mathbf{T}'$ for internal links. In what follows, we shall make these requirements more precise. To this end, let us introduce the relation $\mathbf{AE(T)}$ of an admissible **T**-extension among structures:

Explanation 5.3 (Admissible T-extension) We say that a structure x is an admissible **T**-extension of a structure y and write $(x, y) \in \mathbf{AE(T)}$ if and only if

structures that are obtained by restricting the domain of interpretation of a given structure. In the present case, by contrast, substructure (intended application) and complete structure (**T**-extension) must agree on the domain of interpretation.

(i) x is a **T**-extension of y, (ii) x is a model of **T**, (iii) y is an intended application of **T**, (iv) x satisfies all links to the admissible **T**′-extensions ($\mathbf{T}' \in \mathbf{N}$) of other intended applications and (v) all intended applications of all $\mathbf{T} \in \mathbf{N}$ have an admissible **T**-extension.

Now, our strategy will be to determine the meaning of this relation in a more formal way by means of Carnapian postulates. This means, the relation **AE**(**T**) itself shall be introduced as a theoretical concept. Some postulates for **AE**(**T**) can be advanced by formalising the conditions of Explanation 5.3 in a straightforward manner. Condition (iv) of this explanation, however, requires a few more considerations to be formalised because the satisfaction of links concerns not only the membership of a single pair (x, y) of structures in **AE**(**T**) but the extension of this relation as a whole. Condition (v) will become intelligible when its corresponding postulate is introduced.

The first condition of Explanation 5.3 is rather trivial. An admissible **T**-extension of y must be a **T**-extension of y:

$$\forall x \forall y((x, y) \in \mathbf{AE}(\mathbf{T}) \rightarrow x \in EXT(\mathbf{T})(y)). \qquad \textbf{(P1(T))}$$

The function $EXT(\mathbf{T})$ was defined above by Definition 5.2.

Second, an admissible **T**-extension x needs to satisfy **T**'s substantial law in the sense of being a model of **T**:

$$\forall x \forall y((x, y) \in \mathbf{AE}(\mathbf{T}) \rightarrow x \in \mathbf{M}(\mathbf{T})). \qquad \textbf{(P2(T))}$$

Third, it was required that the range of **AE**(**T**) consists only of intended applications of **T**:

$$\forall x \forall y((x, y) \in \mathbf{AE}(\mathbf{T}) \rightarrow y \in \mathbf{I}(\mathbf{T})). \qquad \textbf{(P3(T))}$$

Condition (iv) of Explanation 5.3 is a bit more involved. As was explained in Sect. 2.3, the need for internal and external links arises when we have overlapping intended applications in the sense that one and the same empirical object is a component of two different intended applications. The most simple example of a link is an equality link which requires that a time-independent function has the same value for one and the same object when this object occurs in two different intended applications. For example, in **LP**, if we put an object a on a balance twice in combination with different other objects, we have two different intended applications y_1 and y_2 such that the admissible **LP**-extensions of y_1 and y_2 are required to agree on the value of the mass function for the argument a. Let us spell out this simple example in a more formal way in order to approach the general form of postulates for links:

$$\forall x_1 \forall y_1 \forall x_2 \forall y_2((x_1, y_1) \in \mathbf{AE}(\mathbf{LP}) \wedge (x_2, y_2) \in \mathbf{AE}(\mathbf{LP}) \rightarrow \forall z(z \in D_{x_1} \cap D_{x_2} \rightarrow t_{x_1}(z) = t_{x_2}(z))).$$

In this notation, D_x designates the base set D of a structure x and $f_x(z)$ the value of a function f in a structure x for an argument z. These notations thus refer to the vocabulary with which the potential and actual models of the corresponding theory-element are formulated.

Equality links may be needed not only to characterise relations among admissible **T**-extensions of intended applications of a single theory-element but also such relations for intended applications of different theory-elements. For example, if an intended application of **LP** is overlapping with one of **CCM** (classical collision mechanics, cf. Sect. 2.3), the admissible **T**-extensions of these intended applications must agree on the values of the mass function in quite the same way as just indicated:

$$\forall x_1 \forall y_1 \forall x_2 \forall y_2 ((x_1, y_1) \in \mathbf{AE(LP)} \wedge (x_2, y_2) \in \mathbf{AE(CCM)} \rightarrow$$

$$\forall z (z \in D_{x_1} \cap P_{x_2} \rightarrow m_{x_1}(z) = m_{x_2}(z))).$$

The distinction between internal and external links goes as follows: internal links concern the **T**-extensions of intended applications of one and the same theory-element, while external links concern the **T**-extensions of intended applications of different theory-elements. External links are particularly important for the determination of intended applications in the following sense. Assume x is an intended application of $\mathbf{T}_i$. Accordingly, the relation concepts $R_1 \ldots R_n$ received a partial interpretation that represents our presumed knowledge, or beliefs, about the empirical system x. This knowledge is, in many cases, not gained through "direct" observation, i.e., observation unaided by any further theory. Rather, axioms of theory-elements different from $\mathbf{T}_i$ are usually in play for the determination of the values of intended applications. Even in our simple example **LP**, where the function n is usually determined for intended applications, measurement of the distance of an object from the centre of rotation is not a matter of direct observation but rests on axioms that introduce a metric among spatial points. Thus, in a structuralist reconstruction the determination of the distance function in **LP** would be represented by an external link between **LP** and another theory-element that accounts for the measurement of such distances.

Internal links need not be binary ones in the sense of concerning the **T**-extensions of exactly two intended applications. A case in point are *extensivity links*. The extensivity link for mass in **LP** requires that, in case a particle a_3 results from a concatenation of two particles a_1 and a_2 (in symbols: $a_3 = a_1 \circ a_2$), then $t(a_3) = t(a_1) + t(a_2)$ must hold. That is, the mass of the concatenated particle must be equivalent with the sum of the masses of the component particles. This link is expressed as follows:

$$\forall x_1 \forall y_1 \forall x_2 \forall y_2 \forall x_3 \forall y_3 ((x_1, y_1) \in \mathbf{AE(LP)} \wedge (x_2, y_2) \in \mathbf{AE(LP)} \wedge (x_3, y_3) \in$$

$$\mathbf{AE(LP)} \rightarrow \forall z_1 \forall z_2 \forall z_3 (z_1 \circ z_2 = z_3 \rightarrow t_{x_1}(z_1) + t_{x_2}(z_2) = t_{x_3}(z_3))).$$

External links are always assumed to be binary in classical structuralism. This assumption is justified by a number of case studies in which binary external links were found sufficiently expressive. Hence, it is only internal links that may have an arity greater than two.

However, there is a way to circumvent non-binary links. Extensivity of a link may simply be considered as the substantial law of another theory-element. That is, in any theory-net, we have, for any extensive quantity, a theory-element that encodes extensivity of this quantity. We see no reason why this transformation of a non-binary link to a theory-element should not work in other cases as well. Notably, equality links, which are binary, are not transformable to theory-elements since for such links it is essential that they concern different intended applications. Henceforth, we assume that that internal and external links are always binary. The motivation for circumventing non-binary links will become obvious when we set out to develop a truth maintenance system with the structuralist approach.[2]

What is the logical form of postulates for (internal and external) links? In the examples discussed, we can distinguish between an antecedent saying that some structures are admissible **T**-extensions of some intended applications and a consequent saying that these admissible **T**-extensions satisfy a certain formula. Hence, we advance the following postulate schema for links:

$$\begin{aligned}&\forall x_1 \forall y_1 \forall x_2 \forall y_2 ((x_1, y_1) \in \mathbf{AE}(\mathbf{T}_j) \wedge \\ &(x_2, y_2) \in \mathbf{AE}(\mathbf{T}_k) \rightarrow \phi(\mathbf{L}_i)(x_1, x_2))\end{aligned} \qquad (\mathbf{P4}(\mathbf{L}_i, \mathbf{T}_j, \mathbf{T}_k))$$

where $\mathbf{T}_j$ and $\mathbf{T}_k$ need not stand for different theory-elements. In the case of an internal link it holds that $\mathbf{T}_j = \mathbf{T}_k$. The adequacy of this postulate schema can be confirmed by studying further structuralist reconstructions of scientific theories. This schema is not solely based on the few examples discussed here.

For the satisfaction of the substantial law, definite postulates could be given using well-defined sets, such as $\mathbf{M}(\mathbf{T})$ and $\mathbf{M_p}(\mathbf{T})$, and well-defined operations on sets, such as $EXT(\mathbf{T})(y)$. As for the satisfaction of links, we must be content with a postulate schema insofar as $\phi(\mathbf{L}_i)(x_1, x_2)$ is a mere placeholder for a specific condition.

Let us move on to specialisation links. The underlying intuition is not difficult to explain: if applying a theory-element $\mathbf{T}'$ to an empirical system represented by a structure y somehow presupposes that another theory-element $\mathbf{T}$ also successfully applies to y, then $\mathbf{T}'$ is considered a specialisation of $\mathbf{T}$. More precisely, $\mathbf{T}'$ being a specialisation of $\mathbf{T}$ – in symbols: $\sigma(\mathbf{T}', \mathbf{T})$ – means that (i) all intended applications of $\mathbf{T}'$ are also intended applications of $\mathbf{T}$, and (ii) for any intended application y of

[2] The price one has to pay for expressing extensivity of a quantity by means of a theory-element as opposed to a link concerns merely the aesthetics of theory-nets. With the introduction of extensivity theory-elements for quantities, certain theory-nets, such as that of classical mechanics, will loose their nicely looking tree structure.

$\mathbf{T}'$, there is a **T**-extension x such that x satisfies the substantial laws of $\mathbf{T}'$ and **T** as well as all the links of these theory-elements. These two requirements are captured by the following postulate:

$$\forall x \forall y((x, y) \in \mathbf{AE}(\mathbf{T}') \rightarrow (x, y) \in \mathbf{AE}(\mathbf{T})). \qquad (\mathbf{P5}(\mathbf{T}, \mathbf{T}'))$$

Such a postulate may be advanced for any two theory-elements $\mathbf{T}'$ and **T** such that $\sigma(\mathbf{T}', \mathbf{T})$. If a single theory-element is considered in isolation, the extension of σ will be the empty set. In this case, no postulate for a specialisation needs to be advanced, of course. As with the postulates for links, we can give only a postulate schema in the form of $\mathbf{P5}(\mathbf{T}, \mathbf{T}')$ for specialisations.

Specialisation was exemplified in Sect. 2.3.4 with classical collision mechanics. There, the law of conservation of energy, which only applies to elastic collisions, was found to be a specialisation of the law of conservation of momentum, which applies to all systems of collision mechanics. So, **ECCM** (elastic classical collision mechanics) is a specialisation of **CCM** (classical collision mechanics). Another specialisation is that between **ICCM** (inelastic classical collision mechanics) and **CCM**. In order to refer to a relation of specialisation, we have also been speaking of a subordinated and a superordinated theory-element.

Now we have achieved a formal translation of the conditions that Explanation 5.3 specifies for a structure x to be an admissible **T**-extension of a structure y. Up to now, the postulates do not express any empirical claim concerning the intended applications of **T**. It might just be the case that no intended application of **T** is successful in the sense of having an admissible **T**-extension. In this case, the extension of $\mathbf{AE}(\mathbf{T})$ would be the empty set. Carnapian postulates, however, also have the function of expressing some empirical content, in addition to specifying the meaning of theoretical concepts. Therefore, we shall advance a postulate that specifies something like a condition of success, i.e., the condition that **T** successfully applies to any of its intended applications. Such an application is successful if only if it can be extended to a **T**-theoretical structure that is a model of **T** and, moreover, satisfies all the links and specialisations to other intended applications. Hence, any intended application of **T** must have an admissible **T**-extension:

$$\forall y(y \in \mathbf{I}(\mathbf{T}) \rightarrow \exists x((x, y) \in \mathbf{AE}(\mathbf{T}))). \qquad (\mathbf{P6}(\mathbf{T}))$$

The postulates $\mathbf{P1}(\mathbf{T})$–$\mathbf{P6}(\mathbf{T})$ do, we claim, completely capture the intended meaning of $\mathbf{AE}(\mathbf{T})$. Henceforth, the set of these postulates is designated by $P(\mathbf{T})$ and is referred to as the *system of postulates for the theory-element* **T**. $D(\mathbf{T})$ designate the set of definitions of $\mathbf{M}(\mathbf{T})$, $\mathbf{M_p}(\mathbf{T})$ and $\mathbf{EXT}(\mathbf{T})$. In the case of a theory-net **N**, $P(\mathbf{N})$ designates the set that contains $\mathbf{P1}(\mathbf{T}) - \mathbf{P6}(\mathbf{T})$ for all $\mathbf{T} \in \mathbf{N}$. Likewise, $D(\mathbf{N})$ designates the set that contains precisely the definitions of $\mathbf{M}(\mathbf{T})$, $\mathbf{M_p}(\mathbf{T})$ and $\mathbf{EXT}(\mathbf{T})$ for all theory-elements in **N**.

5.5 Semantics of Theoretical Terms

5.5.1 Foundations

Let us now spell out the semantics of **AE(T)** and $P(\mathbf{N})$. Following Carnap [33], we shall understand the notion of a theoretical term in such a manner that a theoretical term is a symbol whose meaning is determined through the axioms of a scientific theory. What is it for the meaning of a term to be determined by axioms? Carnap's *Foundations of Logic and Mathematics* [31] contains the seeds of a precise semantics of theoretical terms. He there develops the idea that abstract, or theoretical, terms are *indirectly interpreted* through axioms of a scientific theory. An interpretation of a symbol is indirect if and only if it is given by one or several sentences of the object language as opposed to an assignment of an extension or intension in the metalanguage. Sentences indirectly interpreting symbols are to be adopted as axioms in the calculus of the theory in question. Only observation terms are directly interpreted to the effect that only propositions of the observation language have a direct factual content.

Carnap's notion of indirect interpretation in the *Foundations of Logic and Mathematics* [31] has a strong formalist flavour. However, in his "Beobachtungssprache und theoretische Sprache" [33] we can recognise the elements of a model-theoretic notion of indirect interpretation. There he proposes to divide a scientific theory TC into the Ramsey sentence TC^R and the following conditional:

$$TC^R \rightarrow TC.$$

This conditional became later on labelled the *Carnap sentence* of a scientific theory. Recall that TC designates some conjunction of the T- and C-postulates. T-postulates are those axioms of the theory that contain theoretical terms but no observational ones, whereas C-postulates have occurrences of both theoretical and observation terms. The latter postulates establish connections between the theoretical and the observational vocabulary of the respective scientific theory. Let V_o designate the set of observation terms and V_t the set of theoretical terms. $L(V_o)$ and $L(V_o, V_T)$ designate the respective languages. P_{TC} denotes the set of T- and C-postulates.

The Ramsey sentence of a theory TC in the language $L(V_o, V_t)$ is obtained by the following two transformations of the conjunction of T- and C-postulates. First, replace all theoretical symbols in this conjunction with higher-order variables of appropriate type. Then, bind these variables by higher-order existential quantifiers. As result one obtains a higher-order sentence of the following form:

$$\exists X_1 \ldots \exists X_n TC(n_1, \ldots, n_k, X_1, \ldots, X_n) \qquad (TC^R)$$

where $X_1, \ldots, X_n$ are higher-order variables.

Now, Carnap instructs us to understand the Carnap sentence $TC^R \rightarrow TC$ as follows: if there is an interpretation of the theoretical terms that satisfies TC (in the context of the given interpretation of $L(V_o)$), the theoretical terms should be understood as designating such an interpretation. Notably, the Carnap sentence indirectly interprets the theoretical terms as their interpretation is characterised by the condition that $TC^R \rightarrow TC$ is always true. Carnap [33], however, does not address the problem that $TC^R \rightarrow TC$ usually does not uniquely interpret the theoretical terms. Nor does he consider the case where the Ramsey sentence is false.[3]

Taking up Carnap's train of thought, we can form the notion of an *admissible structure* of the complete language $L(V_o, V_t)$: an $L(V_o, V_t)$ structure is admissible if and only if it (i) extends the given $L(V_o)$ interpretation to interpret the V_t terms and (ii) satisfies the postulates P_{TC}, provided there is such a structure. If there is no such structure, all extensions of $L(V_o)$ that interpret the V_t terms may be considered admissible. $L(V_o, V_t)$ structures may interpret V_t terms in a theoretical domain D_t that expands the observation domain D_o. Henceforth, let $\mathcal{A}_o$ designate the given, or intended, interpretation of the observation language. $EXT(\mathcal{A}_o, V_t, D_t)$ is the set of extensions of $\mathcal{A}_o$ that interpret V_t, where V_t terms may be interpreted in $D_t \cup D_o$. $MOD(A)$ designates the set of models of A, where A is a set of sentences. To make these ideas about admissible structures precise [5]:

Definition 5.3 (Admissible structures) $\mathbf{S}_a$ designates the set of $L(V_o, V_t)$ structures that are admissible by way of an interpretation of the V_t symbols through the postulates P_{TC}.

$$\mathbf{S}_a = \begin{cases} MOD(P_{TC}) \cap EXT(\mathcal{A}_o, V_t, D_t) & \\ & \text{if } MOD(P_{TC}) \cap EXT(\mathcal{A}_o, V_t, D_t) \neq \emptyset \\ EXT(\mathcal{A}_o, V_t, D_t) & \text{if } MOD(P_{TC}) \cap EXT(\mathcal{A}_o, V_t, D_t) = \emptyset. \end{cases}$$

By thus understanding the notion of an admissible structure, the following truth rules are intuitive:

Definition 5.4 (Truth rules for theoretical sentences)
ν: $L(V_o, V_t)^S \rightarrow \{T, F, I\}$, where $L(V_o, V_t)^S$ designates the set of $L(V_o, V_t)$ sentences:

(1) $\nu(\phi) = T$ if and only if for every structure $\mathcal{A} \in \mathbf{S}_a$, $\mathcal{A} \models \phi$
(2) $\nu(\phi) = F$ if and only if for every structure $\mathcal{A} \in \mathbf{S}_a$, $\mathcal{A} \not\models \phi$
(3) $\nu(\phi) = I$ (indeterminate) if and only if there are $\mathcal{A}_1, \mathcal{A}_2 \in \mathbf{S}_a$ such that $\mathcal{A}_1 \models \phi$ and $\mathcal{A}_2 \not\models \phi$.

The underlying idea of these rules is rather simple. A theoretical sentence is true if and only if it is true in every admissible structure. A theoretical sentence is

[3]The former problem is addressed in Carnap [34], where he uses Hilbert's ϵ-operator for expressing that the Carnap sentence yields only an indefinite description of theoretical terms. For a translation of the present semantics of theoretical terms into the language of the epsilon operator, see Andreas and Schiemer [12].

false if and only if it is false in every admissible structure. And a sentence has no determinate truth value if and only if it is true in, at least, one admissible structure and false in, at least, another structure being also admissible. We thus define the notion of truth of a theoretical sentence along the lines of *super truth* in supervaluation semantics. For the claim that a proposition ϕ in a language $L(V_o, V_t)$ is true, we adopt the following notation:

$$\mathbf{S}_a \models_m \phi.$$

So, $\mathbf{S}_a \models_m \phi$ means that, for all structures $\mathcal{A} \in \mathbf{S}_a$ it holds that $\mathcal{A} \models \phi$. The index 'm' stands for 'modal' since the truth rules of Definition 5.4, which underlie this notation, are modal in spirit. It is obvious that in place of $\mathbf{S}_a \models_m \phi$, we could also write $\mathfrak{M} \models \Box\phi$, where $\mathfrak{M}$ is a modal structure in which the set of possible worlds is given by $\mathbf{S}_a$ and every world is accessible from any other and itself.

The semantics of theoretical terms and sentences differs from standard truth-conditional semantics in there being a range of intended interpretations as opposed to there being a single unique one. This requires an adjustment of the notion of logical consequence:

Definition 5.5 ($\Gamma \models_m \phi$) Let Γ be a set of sentences and ϕ be a sentence in a language containing theoretical terms. Then, ϕ is a logical consequence of Γ – in symbols: $\Gamma \models_m \phi$ – if and only if it holds that $\mathbf{S}_a \models_m \phi$ whenever $\mathbf{S}_a \models_m \Gamma$ for any set $\mathbf{S}_a$ of admissible structures.

Fortunately, the peculiar semantics of postulates and theoretical terms requires no modification of the deductive calculus in use because results concerning completeness and soundness obtained for standard semantics remain valid in this semantics. This claim is based on the following proposition [5]:

Proposition 5.1 *For all sets* Γ *of* $L(V_o, V_t)$ *sentences and all sentences* ϕ *of* $L(V_o, V_t)$*:* $\Gamma \models_m \phi$ *if and only if* $\Gamma \models \phi$.

See Appendix for proof. By Proposition 5.1 we know that once a calculus with certain soundness and completeness properties with respect to standard semantics is adopted for a logical language with theoretical terms, these soundness and completeness properties of the calculus will be retained for the present semantics of theoretical terms. That is, if a given deductive system is sound and complete on the standard semantics, this system remains sound and complete on the present theoretical terms semantics.

5.5.2 *Defined Terms*

The semantics of defined terms can be considered as a special case of theoretical terms. For a set V_d of defined terms and a set D of corresponding definitions, $MOD(D) \cap EXT(\mathcal{A}_o, V_t, D_t)$ will be a singleton. Hence Definition 5.3 (which defines the interpretation of theoretical terms) remains applicable to the interpretation of defined terms by corresponding definitions. We merely need to replace the set P_{TC} of postulates with the definitions interpreting the defined terms.

Things are slightly more complicated if the language of our theory TC contains both defined and theoretical terms. Then, the definition of admissible structures needs to be generalised as follows:

Definition 5.6 (Admissible structures of $L(V_o, V_d, V_t)$) $\mathbf{S}_a$ designates the set of $L(V_o, V_d, V_t)$ structures that are admissible by way of an interpretation of the V_t symbols through the postulates P_{TC} and by way of an interpretation of the V_d symbols through the definitions D_{TC}. It is defined as follows:

$$\mathbf{S}_a = \begin{cases} MOD(P_{TC} \cup D_{TC}) \cap EXT(\mathcal{A}_o, V_t \cup V_d, D_t) \\ \quad \text{if } MOD(P_{TC} \cup D_{TC}) \cap EXT(\mathcal{A}_o, V_t \cup V_d, D_t) \neq \emptyset, \\ EXT(\mathcal{A}_o, V_t \cup V_d, D_t) \cap MOD(D_{TC}) \text{ otherwise.} \end{cases}$$

The present truth rules apply nicely to the postulates themselves. If $MOD(P_{TC}) \cap EXT(\mathcal{A}_o, V_t, D_t)$ is not empty, all postulates will be assigned the value true. If the intersection of $MOD(P_{TC})$ and $EXT(\mathcal{A}_o, V_t, D_t)$ turns out to be empty, their truth value will be indeterminate, apart from some contrived cases.[4] In any case, at least one postulate will not be assigned the value true if $MOD(P_{TC}) \cap EXT(\mathcal{A}_o, V_t, D_t)$ is empty. We can say that the valuation of the postulates captures the *empirical claim* of the corresponding scientific theory:

Explanation 5.4 (Empirical claim of a theory) Let Θ be a scientific theory given by the postulates P_{TC}. Then, Θ is empirically true if and only if, for all $\phi \in P_{TC}$, $\nu(\phi) = T$.

This explanation rests on the following considerations. First, the Ramsey sentence TC^R is taken to represent the empirical content of an axiomatic system P_{TC}, following Carnap [33]. Second:

Proposition 5.2 *Let TC^R designate the sentence that is obtained from TC by Ramsification, where TC is interpreted in the domain $\langle D_o, D_t\rangle$. Then, TC^R if and only if $MOD(P_{TC}) \cap EXT(\mathcal{A}_o, V_t, D_t) \neq \emptyset$.*

For proof, see Appendix. And third:

Proposition 5.3 *For all $\phi \in P_{TC}$, $\nu(\phi) = T$ if and only if $MOD(P_{TC}) \cap EXT(\mathcal{A}_o, V_t, D_t) \neq \emptyset$.*

The forward direction of this proposition is trivial to prove. The other direction can easily be proved by contradiction. In sum, Explanation 5.4 follows from the Propositions 5.2 and 5.3, and taking TC^R to express the empirical content of TC.

[4]Here is an example of such a contrived postulate: $\alpha \rightarrow \beta$ with α being an $L(V_o)$ sentence and being falsified by $\mathcal{A}_o$ and β being a theoretical sentence.

5.5.3 *The Semantics of* AE(T)

Let us now spell out the semantics **AE(T)**, the central theoretical term of our system $P(\mathbf{N})$ of postulates. In order to apply the present semantics of theoretical terms to the above axiomatisation of **AE(T)**, we merely need to specify V_o, V_d, V_t, P_{TC}, D_{TC} and D_t in Definition 5.6 for the language $L(\mathbf{N})$, in which the postulates $P(\mathbf{N})$ and definitions $D(\mathbf{N})$ are given. As for V_o, it seems obvious that

$$\ulcorner \mathbf{I}(\mathbf{T}) \urcorner \in V_o(\mathbf{N}) \text{ for all } \mathbf{T} \in \mathbf{N}$$

since $\mathbf{I}(\mathbf{T})$ is assumed to be completely and directly interpreted in structuralism. We do not need to use axioms of $P(\mathbf{N}) \cup D(\mathbf{N})$ to measure $\mathbf{I}(\mathbf{T})$. V_o, furthermore, may or may not contain individual constants designating empirical objects. The observation language of a theory-net **N**, however, does not contain symbols for **T**-non-theoretical concepts since the interpretation of these concepts is specified by the respective components of the intended applications of **T**. Likewise, V_t does not contain **T**-theoretical concepts.

Of V_t we can say that

$$\ulcorner \mathbf{AE}(\mathbf{T}) \urcorner \in V_t(\mathbf{N}) \text{ for all } \mathbf{T} \in \mathbf{N}.$$

$\ulcorner \mathbf{AE}(\mathbf{T}) \urcorner$ is a theoretical term whose meaning is introduced through the postulates $P(\mathbf{N})$. V_t may or may not contain constants designating theoretical objects. V_t does not contain symbols for **T**-theoretical concepts for the same reasons why $V_o(\mathbf{N})$ does not contain symbols for **T**-non-theoretical concepts.

As for P_{TC}, we simply set

$$P_{TC} = P(\mathbf{N})$$

where $P(\mathbf{N})$ is the set of postulates for the relation **AE(T)**, for all $\mathbf{T} \in \mathbf{N}$.

Besides theoretical terms, the language of the present axiom system contains the defined terms $\ulcorner \mathbf{M}(\mathbf{T}) \urcorner$, $\ulcorner \mathbf{M}_p(\mathbf{T}) \urcorner$ and $\ulcorner EXT(\mathbf{T}) \urcorner$. These are set theoretical predicates, which are introduced through explicit definitions. Hence,

$$\ulcorner \mathbf{M}(\mathbf{T}) \urcorner, \ulcorner \mathbf{M}_p(\mathbf{T}) \urcorner, \ulcorner EXT(\mathbf{T}) \urcorner \in V_d(\mathbf{N})$$

$$D_{TC} = D(\mathbf{N}).$$

As all domains of interpretation of **T**-theoretical concepts are given by the intended applications of **T**, no separate theoretical domain of interpretation needs to be specified. Hence, $D_t(\mathbf{N})$ can be set to the empty set. All concepts of the present semantics of theoretical terms have thus been specified for the axiomatic system $P(\mathbf{N}) \cup D(\mathbf{N})$. We can thus define the set of admissible structures of $L(\mathbf{N})$:

Definition 5.7 (Admissible structures of $L(\mathbf{N})$) $\mathbf{S}_a$ designates the set of $L(\mathbf{N})$ structures that are admissible by way of an interpretation of the $V_t(\mathbf{N})$ symbols through the postulates $P(\mathbf{N})$ and by way of an interpretation of the $V_d(\mathbf{N})$ symbols through the definitions $D(\mathbf{N})$.

$$\mathbf{S}_a = \begin{cases} MOD(P(\mathbf{N}) \cup D(\mathbf{N})) \cap EXT(\mathcal{A}_o, V_t(\mathbf{N}) \cup V_d(\mathbf{N}), D_t(\mathbf{N})) \\ \qquad \text{if } MOD(P(\mathbf{N}) \cup D(\mathbf{N})) \cap EXT(\mathcal{A}_o, V_t(\mathbf{N}) \cup V_d(\mathbf{N}), D_t(\mathbf{N})) \neq \emptyset, \\ EXT(\mathcal{A}_o, V_t(\mathbf{N}) \cup V_d(\mathbf{N}), D_t(\mathbf{N})) \cap MOD(D(\mathbf{N})) \text{ otherwise.} \end{cases}$$

Together with the present semantics of theoretical terms, the system $P(\mathbf{N})$ of postulates allows us to account for deductive reasoning within the structuralist framework. (This has been demonstrated in greater detail by Andreas [8].) The present account of structuralism thus axiomatises deductive reasoning with certain types of frame concepts, such as the concepts of an intended application. In the next chapter, we generalise this account to include defeasible forms of reasoning and belief revision.

Chapter 6
Structuralist Belief Revision

Having introduced a system of postulates for structuralism that allows us to represent deductive reasoning in science, we move on to defeasible reasoning and belief changes in the structuralist framework. More precisely, we shall define belief changes in the structuralist representation scheme using the inference system of preferred subtheories explained in Chap. 4.

6.1 Toward a Default Theory for Structuralism

6.1.1 Abductive Reasoning in Science

Belief changes are omnipresent in science. Ptolemy believed that Earth is at rest and located at the centre of our solar system, a belief that was revised by Copernicus, Kepler and Newton. Classical mechanics assumes an Euclidian geometry of space-time, a convention that became revised in Einstein's general theory of relativity. Besides such tremendous and revolutionary changes, which are described as being essentially conceptual ones, there are abundant minor changes in normal science, such as refinements of measurements and revisions of particular hypotheses. Science is by no means a strictly cumulative activity.

Why does science not simply accumulate presumed truths about the world? There is a surprisingly simple answer to this question: scientific reasoning is driven by various kinds of abductive inferences. Such inferences are ampliative in the sense that the conclusion has a richer content than the set of premises. Novel empirical findings and newly advanced scientific theories, therefore, may contradict the conclusion of an abductive inference. In this case, the conclusion may be retracted, particularly if it contradicts empirical findings. Recall from Chap. 3 that any ampliative inference is a defeasible one in the sense that an extension of the premises may invalidate it.

H. Andreas, *Dynamic Tractable Reasoning*, Synthese Library 420,
https://doi.org/10.1007/978-3-030-36233-1_6

Surprisingly, our system $P(\mathbf{N})$ of postulates accounts for simple abductive inferences from phenomena to theoretical descriptions thereof, provided we have beliefs as to which phenomena are to be explained by which theories. It does so by the postulate

$$\forall y(y \in \mathbf{I(T)} \rightarrow \exists x((x, y) \in \mathbf{AE(T)})). \qquad \mathbf{(P6(T))}$$

By universal elimination, this postulate lets us infer that a given intended application b has a theoretical admissible extension a. Using the other postulates, further properties can inferred about this theoretical extension a. In particular, satisfaction of links to the admissible theoretical extensions of other intended applications is ensured by the postulate **P4(T)**. Hence, we can observe that our original system of postulates **P1(T)**–**P5(T)** already allows us to make abductive inferences, albeit expressed in a deductive fashion. Paradoxical as this sounds, the representation of abductive inferences by deductive ones is feasible because the notion of an intended application pairs empirical facts with potential explanations thereof.

Yet, we can also use the structuralist framework to represent the application of generalisations, independently of whether these generalisations are explanatory or not. (By a generalisation we simply mean a universal proposition that may or may not be universally valid.) When we tackle the problem of tractable reasoning in the final two chapters, we exploit the fact that the structuralist system $P(\mathbf{N})$ of postulates accounts for the application of generalisations in terms of certain frame concepts. The final approach to tractable belief revision will therefore be based on a structuralist theory of belief revision to be devised in this chapter.

6.1.2 The Belief Base of a Theory-Net N

When we made ourselves familiar with certain principal approaches to belief revision in Chap. 3.1, we have observed some distinctive merits of using belief bases for the representation of epistemic states. This suggests going for finite belief bases as opposed to belief sets. Let us now specify which types of sentence should be in the belief base for a theory-net **N**:

(1) the postulates **P1(T)**–**P6(T)** for any $\mathbf{T} \in \mathbf{N}$
(2) the definitions of **M(T)**, $\mathbf{M}_p\mathbf{T}$ and EXT(**T**) for any $\mathbf{T} \in \mathbf{N}$
(3) any accepted sentences of the form $b \in \mathbf{I(T)}$ for any $\mathbf{T} \in \mathbf{N}$.

How shall we specify the epistemic ranking among these beliefs? As indicated in Sect. 3.3, no general account of how to determine an epistemic priority ordering has been given in the belief revision literature. Hence, we need to come up with one. In what follows, we shall further elaborate Brewka's proposal to work with a reliability ordering for the items in a belief base [23, 24]. The second step, then, is to carry over this account to theory-elements.

6.2 A Reliabilist Account of Epistemic Ranking

6.2.1 A Simple Case Study

A simple case study, which is adopted from Gärdenfors [61, p. 1], may serve to illustrate conceptual limitations of existing accounts of an epistemic ranking and to motivate a novel account that aims to overcome these limitations:

> Oscar used to believe that he had given Victoria a gold ring at their wedding. He had bought their two rings at a jeweller's shop in Casablanca. He thought it was a bargain. The merchant had claimed that the rings were made of 24 carat gold. They certainly looked like gold, but to be on the safe side Oscar had taken the rings to the jeweller next door who has testified to their gold content. However, some time after the wedding, Oscar was repairing his boat and he noticed that the sulphuric acid he was using stained his ring. He remembered from his school chemistry that the only acid that affected gold was aqua regia. Somewhat surprised, he verified that Victoria's ring was also stained by the acid. So Oscar had to revise his beliefs because they entailed an inconsistency.

Which of Oscar's beliefs are revised and which are retained? This question is not difficult to answer:

> He could not deny that the rings were stained. He toyed with the idea that, by accident, he had bought aqua regia rather than sulphuric acid, but he soon gave up this idea. So, because he had greater confidence in what he was taught in chemistry than in his own smartness, Oscar somewhat downheartedly accepted that the rings were not made of gold after all.

Let us try to give a more formal representation of this example. Oscar considers instances of a number of universal propositions to be true in the process of belief formation and revision. We may represent these propositions by means of open formulas:

$$Looks\_Gold(x) \rightarrow Gold(x) \tag{ε_1}$$

$$Jeweller(x) \wedge Sells(x, y) \wedge Testifies\_Gold(x, y) \rightarrow Gold(y) \tag{ε_2}$$

$$Bargain(x) \rightarrow \neg Gold(x) \tag{ε_3}$$

$$\begin{aligned} Jeweller(x) \wedge \neg Sells(x, y) \wedge Testifies\_Gold(x, y) \\ \rightarrow Gold(y) \end{aligned} \tag{ε_4}$$

$$Sulphuric\_Acid(x) \wedge Stains(x, y) \rightarrow \neg Gold(y). \tag{ε_5}$$

Before the event of exposing the ring to sulphuric acid, Oscar holds the following molecular propositions to be true:

$$Ring(c) \wedge Looks\_Gold(c) \tag{ϕ_1}$$

$$Jeweller(a) \wedge Sells(a, c) \wedge Testifies\_Gold(a, c) \tag{ϕ_2}$$

$$Jeweller(b) \wedge \neg Sells(b, c) \wedge Testifies\_Gold(b, c) \tag{ϕ_3}$$

$$Bargain(c) \tag{ϕ_4}$$

$$Gold(c). \tag{ϕ_5}$$

It would be more precise to say that purchasing the ring is a bargain as opposed to saying that ring c itself is a bargain. Since, however, language is full of metonymies, the present formulation may be considered admissible.

When exposing his ring to sulphuric acid, Oscar finds himself forced to revise his beliefs by the following proposition:

$$Sulphuric\_Acid(d) \wedge Stains(d, c). \qquad (\alpha)$$

As for the belief change resulting from this revision, two principal options are considered:

$$\phi_5 \notin K * \alpha, \text{ while } \varepsilon_5 \in K * \alpha$$

$$\varepsilon_5 \notin K * \alpha, \text{ while } \phi_5 \in K * \alpha.$$

Either the ring is not made of gold or sulphuric acid does stain gold after all. We set aside the case where the staining liquid is aqua regia. As the story goes, Oscar gives up the belief that the ring is made of gold. That is, $\phi_5 \notin K * \alpha$, while $\varepsilon_5 \in K * \alpha$. This result is highly intuitive and uncontroversial. How can it be obtained in the formal accounts of belief revision?

Let us start with belief set revision schemes. For entrenchment-based revisions, the result is obtained if and only if the following two conditions obtain:

$$\neg\alpha <_E \neg\alpha \vee \varepsilon_5$$

$$\neg\alpha \not<_E \neg\alpha \vee \phi_5$$

where $<_E$ represents the strict part of the epistemic entrenchment ordering. ($\alpha <_E \beta$ says that β is more entrenched than α.) This follows from the definition of entrenchment-based revisions (Sect. 3.1.4). Even when taking recourse to informal considerations and intuitions, it seems difficult to justify the above conditions on the entrenchment ordering $<_E$. To be more precise, we are not used to comparing the negation of a new epistemic input α with the disjunction of this negation and a previous belief in terms of their epistemic quality. In any case, we lack a general account of determining epistemic entrenchment orderings. Partial meet revisions and the system of spheres approach by Grove [67] fare hardly better than entrenchment-based revisions.

Let us move on to belief base revision schemes. To study these, we need to divide the explicitly stated beliefs of the story into basic ones, derived ones and expectations. Arguably, ϕ_1, ϕ_2 and ϕ_3 are basic beliefs. Also, ϕ_4 is best considered basic since we are not told how Oscar came to hold this belief. ϕ_5, by contrast, is clearly derived. $\varepsilon_1, \ldots, \varepsilon_5$ are expectations. Both basic beliefs (in the narrow sense) and expectations are members of the belief base. How can we specify the epistemic ranking among these beliefs?

6.2.2 *The Reliability of Expectations*

At the time of purchasing the rings, ε_4 is most decisive for Oscar's coming to believe that they are made of gold. When he finds the ring stained by sulphuric acid, however, ε_5 comes into play and overrides ε_4. Oscar gives the result of the simple and accidental chemical test more credit than the justification-based on the testimony of the jewellers. In other words, the chemical test is considered more reliable than the testimony of the jewellers. Why is this? Presumably, there is no case known to Oscar where gold was stained by sulphuric acid, whereas Oscar knows that it happens from time to time that imitations of gold are sold without proper declaration. If so, the reliability order used by Oscar can, at least partially, be explained by the ratio of the number of successful applications to the total number of successful and unsuccessful applications of the expectations in play. We say that an application of an expectation is successful if and only if its conclusion has not been withdrawn so far. Otherwise, it is unsuccessful.

Goldman [66] defines the concept of reliability as the tendency of a belief-forming process to produce beliefs that are true rather than false. The term *tendency* can, Goldman explains, either refer to actual long run frequency of truth versus error or to a propensity determined through the outcomes that would occur in merely possible realisations of the process. These ideas are adopted here, with the qualification that reliability is not evaluated in terms of truth but in terms of acceptance. The more frequent the inferences of an expectation remain accepted, the higher is the reliability of this expectation. For the time being, only actual applications of inferential patterns are considered for the estimation of reliability but not potential ones.

Now, the only ranking information needed to account for Oscar's revision in the story is an order of reliability of the general axioms. For example, the following strict total order among the axioms may be assumed for the story:

$$\varepsilon_1 <_R \varepsilon_2 <_R \varepsilon_3 <_R \varepsilon_4 <_R \varepsilon_5$$

where $\varepsilon_i <_R \varepsilon_j$ means that ε_i is strictly less reliable than ε_j. A strict total order for the reliability of the axioms seems neither necessary nor adequate, however. So it seems more appropriate to say that the following information about the order of axioms is actually used in the story:

$$\varepsilon_1, \varepsilon_2, \varepsilon_3, \varepsilon_4 <_R \varepsilon_5$$

$$\varepsilon_1, \varepsilon_2, \varepsilon_3 <_R \varepsilon_4.$$

This partial order is modular and, hence, can be represented by the levels of a default theory.

Another observation concerning the example can now be made. It appears that no ranking information concerning the non-derived beliefs is needed to account for Oscar's revision considerations. That the ring is stained seems to be accepted as firmly as the sentence saying that Oscar was told that the ring is made of gold by the jeweller. The absence of any prioritisation among non-derived beliefs in the present example may be specified as follows:

$$\phi_i \leq \phi_j \wedge \phi_j \leq \phi_i \text{ for all i,j } (1 \leq i, j \leq 4).$$

We write $\phi_i \leq \phi_j$ as opposed to $\phi_i \leq_R \phi_j$ for ϕ_i, ϕ_j being basic beliefs since, on the present understanding of reliability, it is only belief-forming inferential patterns but not atomic and molecular propositions that are comparable in terms of their reliability. $\phi_i \leq \phi_j$ means that ϕ_i has at least the same epistemic priority as ϕ_j, $\phi_i < \phi_j$ that ϕ_i has higher epistemic priority than ϕ_j.

To generalise the observation about the epistemic priority of non-derived beliefs:

Proposition 6.1 *All non-derived beliefs have equal epistemic priority.*

Since $\alpha < \beta$ means that α is epistemically better than β, epistemic priority for expectations is the inverse of their reliability:

Proposition 6.2 *For all expectations $\varepsilon_i, \varepsilon_j$: $\varepsilon_i \leq \varepsilon_j$ if and only if $\varepsilon_j \leq_R \varepsilon_i$.*

There remains to specify the ordering between basic beliefs and expectations. We suggest that a proper basic belief, i.e., a basic belief in the narrow sense, always has priority over an expectation:

Proposition 6.3 *If ϕ_i is a proper basic belief and ε_j an expectation, $\phi_i < \varepsilon_j$.*

Notably, Proposition 6.1 allows us to determine the epistemic priority of a new epistemic input relative to the other items of the belief base. If the new epistemic input is a non-derived belief, it has the epistemic priority of the other non-derived beliefs. If the new epistemic input has the form of a general axiom, its epistemic priority directly depends on its history of successful and unsuccessful applications.

Propositions 6.1 and 6.3 are reminiscent of the foundationalist doctrine that knowledge has infallible foundations (cf. [98, Sect. 5.1]). In fact, they are in conflict with the anti-foundationalist contention that even presumed basic beliefs may be retracted and superseded by other beliefs. A standard example of this is hallucination. In hallucination, however, beliefs about the spatiotemporal world become a subject of theorising and thus cease to be proper basic beliefs. For example, when arguing that some agent was trapped in a hallucination, then, in the logical analysis of this argument, the agent's beliefs about the spatiotemporal world are represented as being derived from sentences about the agent's sensations. Of course, the division between derived and non-derived beliefs is dynamically dependent upon the state of the art of our theories and upon which theories are brought to bear on the analysis of beliefs.[1]

[1] For a criticism of arguments purporting to show that sentences about sensations are semantically dependent upon sentences about external things, see Andreas [4].

Though reminiscent of the doctrine of infallible foundations, Propositions 6.1 and 6.3 do not imply that proper basic beliefs are never retracted. Assume some new epistemic input α is non-derived and inconsistent with some non-derived ϕ, which has been accepted so far. Then, there will be no preferred subtheory containing both α and ϕ. Hence, neither α nor ϕ will be believed by the subject. This amounts to a retraction of ϕ without ϕ becoming superseded by the new epistemic input. Note that reasoning does not break down for inconsistent, non-prioritised bases since the inference formalism of PSA is paraconsistent.

The belief base H, the set E of expectations and their reliability ordering have now been specified to an extent that the definitions of revisions in the PSA scheme can be applied. It is easy to verify that applying these definitions yields precisely the results that are reported in the story about Oscar's ring. Since ε_5 has the highest priority among the expectations, once the accidental test with sulphuric acid has been conducted, the proposition that the ring is made of gold cannot be included in any preferred subtheory. Before the ring was stained, Oscar accepted the proposition in question because of ε_4 and because ε_4 has priority over ε_3. No prioritisation among the non-derived beliefs $\phi_1, \ldots, \phi_4$ is recognisable in the story.

To sum up some, still very conjectural results concerning the epistemic ranking of a belief base and its revision:

(1) No epistemic ranking of non-derived beliefs is needed to account for the revision of a ranked belief base.
(2) No explicit epistemic ranking of derived beliefs needs to be given to account for revisions of such beliefs. Rather, conflicts at the level of derived beliefs can be resolved by considering the inferential patterns that were effective for the acceptance of conflicting derived beliefs. Only these inferential patterns need to have a ranking.
(3) The ranking of inferential patterns can, at least in part, be explained in terms of statistical concepts, viz., by the frequency of their successful application. The higher this frequency is, the more reliable is the inferential pattern.

6.2.3 *The Epistemic Ranking of Theory-Elements*

Let us resume the discussion of revising intended applications of theory-elements in structuralism. Arguably, theory-elements are comparable to expectations, i.e., defeasibly valid axioms of background theories in the direct mode of base revisions. More precisely, theory-elements encode axioms whose application to empirical systems leads to the formation of derived beliefs. Such an application is considered defeasible just as expectations are. Furthermore, the distinction between a successful and an unsuccessful application of an inferential pattern makes sense equally for intended applications:

Explanation 6.1 (Successful intended application) An intended application y of a theory-element **T** is successful if and only if **T**'s substantial law has been used to derive a sentence ϕ about the objects of the domain of y, and ϕ has remained accepted so far.

In other words, an intended application is successful if and only if it contributes to the formation of derived beliefs. For example, an experimental test whose result became accepted without later revision is a successful intended application of the corresponding theory-element. If, by contrast, the result became retracted at a later stage of inquiry, this would qualify as an unsuccessful intended application. There may be intended applications which are neither successful nor unsuccessful. Thus, we can introduce a priority ordering among theory-elements in terms of their reliability:

Definition 6.1 (Epistemic priority of theory-elements) A theory-element $\mathbf{T}_i$ has priority over another $\mathbf{T}_j$ if and only if $\mathbf{T}_i$ is more *reliable* than $\mathbf{T}_j$.

As for the inference system that generates a belief set on the basis of a belief base, we have to find one that respects the priority ordering among theory-elements.

A somewhat more extended discussion of reliability is in order here. This concept was introduced with reference to Goldman [66] and the qualification that reliability of an inferential pattern is interpreted as actual long run frequency of successful applications. Success of an inferential pattern in turn is not understood in terms truth conduciveness but in terms of acceptance conduciveness.[2]

A finite frequency interpretation of reliability works fine for a large number of cases and deserves, for this reason, to be made explicit. Let us assume, for now, that the theory-elements have a fairly significant history of applications that is not represented by the sets $\mathbf{I}(\mathbf{T})$. This assumption enables us to evaluate reliability in terms of actual long run frequencies of success. Let $C_s(\mathbf{T})$ denote the cardinality of the set of successful applications of $\mathbf{T}$ and $C_u(\mathbf{T})$ denote the cardinality of the set of unsuccessful applications of $\mathbf{T}$. Then, we can define a priority order as follows:

$$\mathbf{T}_i \leq_R \mathbf{T}_j \text{ iff } \frac{C_s(\mathbf{T}_i)}{C_s(\mathbf{T}_i) + C_u(\mathbf{T}_i)} \leq \frac{C_s(\mathbf{T}_j)}{C_s(\mathbf{T}_j) + C_u(\mathbf{T}_j)} \tag{6.1}$$

where $\mathbf{T}_i \leq_R \mathbf{T}_j$ means that $\mathbf{T}_i$ is at most as reliable as $\mathbf{T}_j$. This definition of reliability, however, appears too restrictive. Consider two theory-elements $\mathbf{T}_i$ and $\mathbf{T}_j$ such that an application of $\mathbf{T}_i$ conflicts with one of $\mathbf{T}_j$ and the statistics of $\mathbf{T}_j$ is only marginally better than that of $\mathbf{T}_i$. In a context where we are forced to choose between $\mathbf{T}_i$ and $\mathbf{T}_j$ quickly, we may simply go for $\mathbf{T}_j$, then. In a scientific context, however, it might not be necessary to make an immediate decision. Then, we may simply wait until further empirical findings or further theoretical developments lend stronger support to the application of either theory-element. In view of such a scenario, one

[2] Arguably, in the examples that Goldman [66] discusses when advancing his reliabilist account of justification, belief-forming processes are assessed merely in terms of long run acceptance rather than in terms of truth. Notably, such a reading would go clearly against the externalist intentions of Goldman.

better defines that $\mathbf{T}_j$ is more reliable than $\mathbf{T}_i$ if and only if the statistics of successful applications of $\mathbf{T}_j$ is significantly better than that of $\mathbf{T}_i$. This may be expressed by requiring the ratio of the statistics of $\mathbf{T}_j$ to that of $\mathbf{T}_i$ to be higher or equal than a certain margin m, $m > 1$:

$$\mathbf{T}_i <_R \mathbf{T}_j \text{ if and only if } \frac{\dfrac{C_s(\mathbf{T}_j)}{C_s(\mathbf{T}_j) + C_u(\mathbf{T}_j)}}{\dfrac{C_s(\mathbf{T}_i)}{C_s(\mathbf{T}_i) + C_u(\mathbf{T}_i)}} \geq m. \tag{6.2}$$

The formal properties of $\leq_R$ and $<_R$ depend on which definition of reliability we adopt and to what extent a statistics of successful applications is available for the theory-elements of **N**. If we take (6.1) to define reliability and there is such a statistics available for all theory-elements, $\leq_R$ will be a total preorder. The strict part of $\leq_R$ (which is obtained by $\mathbf{T}_i <_R \mathbf{T}_j$ if and only if $\mathbf{T}_i \leq_R \mathbf{T}_j$ but not $\mathbf{T}_j \leq_R \mathbf{T}_i$) is a strict partial order that is modular, then. Equation (6.2), by contrast, defines a strict partial order that will not always be modular. If there are theory-elements of which there is no statistics of successful applications available, (6.1) cannot be used to define a total preorder $\leq_R$ of reliability. The strict part of $\leq_R$ will then not be modular.

There is a generalisation of PSA for partially ordered theories the order of which is not modular [23, 24]. The corresponding definition of belief revisions in PSA are easy to adjust to partially ordered theories. For simplicity, however, we have assumed a modular reliability ordering of default theories. Modularity can be achieved by defining a partition $I_1, \ldots, I_n$ of the interval [0, 1] of rational numbers. Then, we can easily define the epistemic ranking as follows:

Definition 6.2 ($<_R$) Let $\langle I_1, \ldots, I_n \rangle$ be a sequence of intervals that forms a partition of the interval [0, 1] of rational numbers. For all $i, j, x, y : x \in I_i, y \in I_j$ and $i < j$ implies that $x < y$. $\mathbf{T}_i <_R \mathbf{T}_j$ if and only if there are k and l such that

$$\frac{C_s(\mathbf{T}_i)}{C_s(\mathbf{T}_i) + C_u(\mathbf{T}_i)} \in I_k, \ \frac{C_s(\mathbf{T}_j)}{C_s(\mathbf{T}_j) + C_u(\mathbf{T}_j)} \in I_l, \text{ and } k < l.$$

Such a partition of the reliability interval between zero and one does not entirely avoid the problem that marginal differences in the success statistics lead to a prioritisation, but it diminishes this problem. For simplicity, we assume that the reliability ordering among theory-elements is modular, while noting that all subsequent definitions of structuralist belief revision can be generalised so as to take non-modular partial reliability orderings into account.[3]

[3]Even though the finite frequency interpretation of reliability can be shown to work well for a large number of cases, it must be admitted that this interpretation may not exhaust our understanding of the concept of reliability. Judgements concerning reliability need not be based exclusively on the statistics of successful applications. Consider the inference from a speaker's assertion of a sentence ϕ to ϕ itself. Here, judgements about the competency of the speaker may well guide the estimation

There remain to address, finally, ideas about hierarchies of laws, which are represented in the structuralist approach by means of the specialisation relation $\sigma(\mathbf{T}', \mathbf{T})$ (Sect. 2.3.4). Gärdenfors [61, p. 92n] sees such a relation as encoding a rough sorting of sentences into different degrees of epistemic entrenchment. The higher up an axiom is in the corresponding hierarchy of theory-elements, the better is its standing in the epistemic entrenchment ordering. We must wonder, therefore, whether we need another metatheoretical rule saying that a theory-element always has priority over any of its specialisations. This sort of hierarchy among theory-elements is respected, to some extent, in the system of postulates:

$$\mathbf{P5}(\mathbf{T}, \mathbf{T}') \quad \forall x \forall y((x, y) \in \mathbf{AE}(\mathbf{T}') \rightarrow (x, y) \in \mathbf{AE}(\mathbf{T})).$$

This postulate says that, for a $\mathbf{T}'$-extension x of y to be admissible, x must also be a $\mathbf{T}$-extension of y if $\mathbf{T}'$ is a specialisation of $\mathbf{T}$. The instances of this schema thus ensure that $(x, y) \in \mathbf{AE}(\mathbf{T}')$ is inferred only on condition of $(x, y) \in \mathbf{AE}(\mathbf{T})$ being consistent with the present set of accepted sentences. Hence, priorities among laws that are expressed by specialisation among theory-elements are taken into account by postulate **P5**.

How does the epistemic ranking of theory-elements translate to an epistemic ranking of the sentences of a structuralist belief base? Recall from Sect. 6.1.2 that a belief base for a theory-net $\mathbf{N}$ contains $P(\mathbf{N}) \cup D(\mathbf{N})$ for any $\mathbf{T} \in \mathbf{N}$ as well as a sentential representation of the intended applications of $\mathbf{N}$. To translate the epistemic ranking of theory-elements into an epistemic ordering of such a belief base, we can simply introduce an epistemic ranking among the sentential representations of the intended applications of $\mathbf{N}$. To be more precise, for any intended applications b_k and b_l:

$$b_k \in \mathbf{I}(\mathbf{T}_i) < b_l \in \mathbf{I}(\mathbf{T}_j) \text{ iff } \mathbf{T}_j <_R \mathbf{T}_i.$$

This has the effect that intended applications of epistemically superior theory-elements have priority in case of a conflict with epistemically inferior intended applications. Such a conflict arises if two intended applications do not have corresponding theoretical admissible extensions such that all links are satisfied.

of the reliability of the inference. For example, knowing that someone holds a chair of theoretical physics suggests that the reliability of his assertions concerning quantum mechanics is particularly high, even in the absence of information about the success statistics of the inferential pattern that allows for the transition from the assertion of ϕ by this person to ϕ itself. Still, from what we know about this person, we can infer that this statistics must be pretty high.

Adopting a strict finite frequency interpretation of reliability is a way of *defining* reliability. Allowing considerations other than the statistics of successful applications to determine judgements about reliability may be described as the view that reliability is a theoretical concept. On this view, (6.1) and (6.2) are not alternative definitions of reliability but alternative C-postulates that introduce the concept of reliability and which may well be complemented by further C- and T-postulates, such as an axiom connecting the reliability of a speaker with her competency. At the present stage of formalisation, these postulates are ones of the metatheory, not of the object theory.

6.3 The Final Synthesis

Where did we get so far on our way to a structuralist approach to belief revision? The result of the preceding and the present chapter may be described as a dynamisation of classical structuralism by means of the following operations: (i) introduction of the relation **AE(T)** by postulates, (ii) definition of a structuralist belief base and (iii) introduction of an epistemic ranking among the theory-elements. The connection to belief revision theory is now easy to establish. We merely need to fill the placeholders in the above explained schema for defining direct base revisions:

$$K *_i \alpha = Inf(\mathcal{H} \oplus_i \alpha)$$

where $\mathcal{H}$ stands for a ranked belief base and Inf for a nonmonotonic and paraconsistent inference operation. Using the distinction between proper basic beliefs and background theory, we can fill in the elements of the schema as follows:

(1) The belief base $H(\mathbf{N})$ contains any accepted proposition saying that a structure b_i is an intended application of a theory-element $\mathbf{T}_j$. Only such propositions are elements of $H(\mathbf{N})$.
(2) The background theory $E(\mathbf{N})$ is given by the set $P(\mathbf{N}) \cup D(\mathbf{N})$.
(3) The inference operation Inf is given by PSA, the preferred subtheories approach to defeasible reasoning.

As we define structuralist belief changes by means of preferred subtheories, we need to define the notion of a structuralist default theory:

Definition 6.3 (Structuralist default theory $\Theta(\mathbf{N})$) $\Theta(\mathbf{N})$ is a structuralist default theory if and only if there are $\mathbf{N}$, $E(\mathbf{N})$, $H(\mathbf{N})$ such that

(1) $\mathbf{N}$ is a theory-net in the sense of structuralism
(2) $\Theta(\mathbf{N}) = \langle E(\mathbf{N}), H_1, \ldots, H_n \rangle$
(3) The subsequence $\langle H_1, \ldots, H_n \rangle$ of $\Theta(\mathbf{N})$ is a partition of $H(\mathbf{N})$ such that for all $\alpha, \beta \in H(\mathbf{N})$, $\alpha = \ulcorner b_i \in \mathbf{I}(\mathbf{T}_j)\urcorner$ and $\beta = \ulcorner b_k \in \mathbf{I}(\mathbf{T}_l)\urcorner$,

    (a) $\alpha \not< \beta$ and $\beta \not< \alpha$ if and only if $\mathbf{T}_j \not<_R \mathbf{T}_l$ and $\mathbf{T}_l \not<_R \mathbf{T}_j$
    (b) $\alpha < \beta$ if and only if $\mathbf{T}_l <_R \mathbf{T}_j$.

Condition (3) of this definition simply says that $H_1, \ldots, H_n$ are equivalence classes of intended applications, where two intended applications are equivalent if and only if their theory-elements have the same reliability. Further, it says that H_1 contains the intended applications of the most reliable theory-elements, H_2 the second most reliable ones, etc. The relation $<_R$ is defined by Definition 6.2. $\alpha < \beta$ means that $\alpha \in H_k$ and $\beta \in H_m$ such that $k < m$.

Now we are in a position to define revisions and contractions along the lines of Brewka [24] and Rott [98]:

Definition 6.4 ($K(\Theta(\mathbf{N}))$ and $K(\Theta(\mathbf{N})) *_i \alpha$) Let $\Theta(\mathbf{N}) = \langle E(\mathbf{N}), H_1, \ldots, H_n \rangle$ be a structuralist theory. α is a sentence of the form $b \in \mathbf{I}(\mathbf{T})$. The formation of the belief set $K(\Theta(\mathbf{N}))$ and revision thereof are defined as follows:

$$Inf(\Theta(\mathbf{N})) = \{\phi \mid \Theta(\mathbf{N}) \mid\!\sim^{PS} \phi\}$$

$$K(\Theta(\mathbf{N})) = Inf(\Theta(\mathbf{N}))$$

$$K(\Theta(\mathbf{N})) *_i \alpha = Inf(\Theta(\mathbf{N}) \oplus_i \alpha)$$

where $\mid\!\sim^{PS}$ is a placeholder to be filled with either $\mid\!\sim^{PS(su)}$ or $\mid\!\sim^{PS(s)}$.

This is the final step in the synthesis of the structuralist framework with belief revision theory in the medium of some system of defeasible reasoning. Recall that postulate **P6**(**T**) makes it a conceptual truth that intended applications are successful in the sense of having admissible **T**-extensions. In the present system, however, the set $H(\mathbf{N}) \cup E(\mathbf{N})$ will usually not be consistent so that some intended applications will not be successful. Hence, the construction of preferred subtheories will decide which intended applications remain accepted as successful. If α is a sentence of the form $b \in I(\mathbf{T})$ and thus a member of $H(\mathbf{N})$ but not a member of any preferred subtheory of $\Theta(\mathbf{N})$, α will be classified as being unsuccessful. In this case, α is not inferable from $\Theta(\mathbf{N})$ in PSA. If α is a member of some but not all preferred subtheories, we consider it partially successful. Finally, α is successful if and only if it is a member of all preferred subtheories.

Chapter 7
Truth Maintenance

It is time to tackle the problem of tractable reasoning: a truth maintenance system (TMS) will be devised on the basis of the structuralist belief revision scheme. We want to show how belief changes are cognitively and computationally feasible. We prove that the TMS achieves a tractable computation of belief changes, which approximates the structuralist belief revision scheme. The study of truth maintenance with frames in this chapter helps us devise frame logic and a novel belief revision scheme in the next chapter.

7.1 Computational Complexity and Cognition

7.1.1 Tractability

What does it mean that belief revision is intractable? In non-technical terms, it means that our best computers take far too much time to compute the revision of "larger" belief systems. Intractable problems are also called *computationally infeasible*. To explain the notion of tractability in more technical terms, a very elementary review of the basic concepts of the theory of computational complexity is necessary.[1]

A major objective of the theory of computational complexity is to classify problems in terms of the running time that a machine takes to solve them. Such a classification must refer to some conception of a universal machine in order to have a common standard for measuring running time. The concept of a Turing machine was found appropriate for that purpose. It is well defined, for such a machine, what the single steps of computation are: reading input, writing output and possibly changing

[1]The following review is mainly based on Arora and Barak [14], which is also recommended for further reference.

H. Andreas, *Dynamic Tractable Reasoning*, Synthese Library 420,
https://doi.org/10.1007/978-3-030-36233-1_7

the state, possibly moving the head one step to the left or to the right. So it seems promising to define a measure of running time (that a computer takes to solve a problem) in terms of the number of computation steps that a Turing machine takes to solve the problem.

The notion of a problem is understood as the computation of the value of a function for a given argument. The domain and the range of such a function is a set of binary strings. This is no serious restriction since strings of finite alphabets can be encoded by ones of a binary alphabet. That there is such an encoding will be tacitly assumed henceforth. It is easy to see that many problems in logic, mathematics, physics, engineering, etc. are captured by this notion of a problem. Think, for example, of a function that takes first-order formulas as input and has the output *true* if the formula is a theorem of some first-order calculus and the output *false* if it is not. Further, the problem of computing the trajectories of the planets of our solar system, given certain boundary conditions, can easily be formulated as computing the values of a function.

Let $f : \{0, 1\}^* \to \{0, 1\}^*$ be a function such that its computation for a given input string generates solutions to a problem. ($\{0, 1\}^*$ designates a set of binary strings.) Assume there is a Turing machine M that computes, for input strings $x \in \{0, 1\}^*$, the value $f(x)$ and halts when the computation has been completed. One says that M computes f in $T(n)$-time if its computation on any input x requires at most $T(|x|)$ steps, where $T : \mathbb{N} \to \mathbb{N}$ and $|x|$ is the length of the input string. In case f is a Boolean function, f defines a decision problem and a language. One says that a machine decides the language L if it computes the function $f_L : \{0, 1\}^* \to \{0, 1\}$, where $f_L(x) = 1$ if and only if $x \in L$.

Even though the concept of a Turing machine serves as universal machine model, one does not want to determine the precise number of computation steps that some Turing machine takes to solve a problem. The idea rather is to distinguish certain classes of functions $T(|x|)$ such that these classes provide a measure of running time. As for the difference between a tractable and an intractable problem, the classes **P**, **NP** and **EXP** are of crucial interest. **P** stands for polynomial time, **NP** for nondeterministic polynomial time and **EXP** for exponential time.

A decision problem belongs to the class **P** if and only if there is a Turing machine that decides it in polynomial time, i.e., there is a constant $c > 0$ such that the computation takes at most n^c steps, where n is the length of the input string. Analogously, a decision problem belongs to the class **EXP** if and only if there is a Turing machine M and a constant $c \geq 1$ such that M decides the problem taking at most 2^{n^c} computation steps.

The explanation of the class **NP** is more intricate. This class is intended to capture those problems whose "solution" can be verified in polynomial time. What a solution of a decision problem is varies from problem to problem. For the problem of deciding whether a formula ϕ of propositional logic is satisfiable, a presumed solution has the form of a valuation of the propositional constants that occur in ϕ. Whether or not such a valuation satisfies ϕ can be decided in polynomial time. Hence, deciding whether a given formula of the propositional calculus is satisfiable is a problem in **NP**. An actual solution of a problem instance is also called a *certificate* of it. In essence, a certificate is some piece of information that allows one to verify membership of the input string in the language under consideration.

An alternative definition of **NP** rests on the notion of a nondeterministic Turing machine that is allowed to make nondeterministic choices. Such choices can be used to encode potential certificates that are verifiable or falsifiable in polynomial time. Hence, **NP** stands for decidable in nondeterministic polynomial time. No actual machine is nondeterministic in this sense, however.

When actually classifying a problem using these measures, one must study algorithms that decide it, provided there is any such algorithm at all. In doing so, it is not necessary to come up with an implementation of the algorithms on a Turing machine. According to the Church-Turing thesis, every physically realisable computational device can be implemented by a Turing machine. Moreover, and this point matters for the use of complexity measures, the strong form of the Church-Turing thesis says that this simulation takes only polynomial overhead, i.e., t computation steps on the given computation model can be simulated by t^c steps on the Turing machine, where c is a constant. For these two reasons, one can determine the complexity class of an algorithm by analysing the steps of computation that would take its execution on some physically realisable computation model, including the *von Neumann architecture*. All digital computers today are realisations of this architecture.

It is important to note, furthermore, that the classification of algorithms into complexity classes concerns their asymptotic behaviour. For example, an algorithm is said to be in the class **P** if and only if there are natural numbers c and n_0, and a polynomial $p(n)$ such that the execution of the algorithm takes, for all $n > n_0$, at most $c\, p(n)$ steps, where n designates the size of the input. Thus, membership of a problem in, say **P**, means (i) that there exists an algorithm that decides it and (ii) the number of computation steps of this algorithm is asymptotically bounded by a polynomial $p(n)$.

Having given an elementary review of these notions in the theory of computational complexity, we are in a position to say what it is for a problem to be tractable. Simply, a problem is called *tractable* if and only if it is in the class **P**. That is, it is tractable if and only if there is an algorithm that decides it in polynomial running time. Algorithms with polynomial running time are also called *efficient*.

Of particular interest are the classes **NP**-complete and **NP**-hard. A decision problem is **NP**-hard if and only if every decision problem in **NP** can be reduced to it. Reduction here means that a decision problem p' can be translated into another p (in polynomial time) such that the decision of p gives the decision of p' at the same time. Thus, **NP**-hardness of a problem means that the problem is at least as difficult as any other **NP**-problem. A problem is **NP**-complete if and only if it is **NP**-hard and a member of **NP**. Satisfiability of a formula in propositional logic is **NP**-complete. For complexity classes other than **NP**, hardness and completeness are defined analogously.

It seems as if membership of a problem in the class **NP**-complete means that the problem is intractable. No proof, however, could be given that **P** and **NP** are distinct classes. Whether or not $\mathbf{P} \neq \mathbf{NP}$ is considered the central open question of complexity theory. Still, it is strongly believed that $\mathbf{P} \neq \mathbf{NP}$. If this is true, **NP**-complete problems escape efficient computation. **EXP**-complete problems are certainly intractable since it can be shown that $\mathbf{P} \subset \mathbf{EXP}$.

All **NP**-complete problems can be decided by algorithms with exponential running time. This is particularly easy to verify for the problem of satisfiability of a formula in propositional logic. For an algorithm to have exponential running time means that the number of computation steps that are needed for a decision grows, in worst-case scenarios, exponentially with the syntactic size of the problem instance. This behaviour is also referred to as the *problem of combinatorial explosion*. Exponential growth of running time is considered a mark of intractability for obvious reasons.

For every decision problem concerning membership of a string in the language L, there is a complementary problem that consists in determining whether a given string is not a member of the language L. The class **coNP** is defined as the class of decision problems for which the complementary problem is in **NP**. Determining whether a given formula in propositional logic is a tautology is a famous **coNP**-complete problem. The classes **coNP** and **P** are strongly believed to be distinct, but no proof could be given so far for this presumption. Likewise, the precise relation between the classes **coNP** and **NP** is an open question.

Suppose some problem is such that its decision requires deciding a subproblem a number of times, where this number depends on the size of the original problem. Further, suppose the subproblem is known to be a member of a distinct complexity class. To take such compositions of problems into account, one hypothetically assumes recourse to an *oracle*, which is capable of solving the subproblem in constant time. The complexity of the composite problem, then, is expressed in the following manner. $\mathbf{P}^{\mathbf{X}}$ denotes the class of problems that can be decided in polynomial time by a deterministic Turing machine that is allowed to use an oracle for deciding a subproblem in the class **X**. $\mathbf{NP}^{\mathbf{X}}$, likewise, denotes the class of problems that can be decided in nondeterministic polynomial time by a Turing machine that is allowed to use oracles for deciding a subproblem in **X**. Using these notions, a hierarchy of complexity classes can be defined that reflects a corresponding hierarchy of problems. The details of this hierarchy need not concern us here, even though they proved relevant in the study of both belief revision and nonmonotonic reasoning.

It goes without saying that intractability is not the worst case we may encounter when dealing with decision problems. As is well known, full first-order logic is undecidable, i.e., there is no algorithm available that determines, on all admissible inputs, whether a given formula is a theorem of first-order logic. One means to achieve decidability in first-order contexts is to constrain the domain of interpretation to finite ones.

7.1.2 Computation of Revisions

The computation of revisions is taken to consist in determining whether a given sentence β is a member of the knowledge set K when revised by some sentence α. Belief revision, in this formulation, is the problem of deciding whether

$$\beta \in K * \alpha$$

for arbitrary formulas α and β of the respective language L.

Syntax-based approaches to belief revision are computationally at least as demanding as the determination of consistency and entailment. For, both problems figure as subproblems in these approaches. As is well known, consistency and entailment can be determined using the test of satisfiability. Deciding the consistency of a finite set of formulas reduces to deciding the satisfiability of a single formula. Since entailment can be reduced to deciding non-satisfiability of a formula, entailment is complementary to satisfiability. Since satisfiability is **NP**-complete, these considerations show that (syntax-based) belief revision in a propositional language cannot be in a complexity class below **NP**. It can even be shown that, provided $\mathbf{P} \neq \mathbf{NP}$, belief revision with belief bases in a propositional language is computationally more demanding than satisfiability and non-satisfiability in propositional logic. It is instructive to see why this is so.

As pointed out by Nebel [89, Chaps. 5.1 and 5.2], the problem of revising a belief base – be it prioritised or not – requires solving two subproblems. First, we need to test, several times, the consistency of subsets of the belief base. Second, we need to solve either (i) an entailment problem for all members of a set of inclusion-maximal subsets of the belief base H or (ii) an entailment problem for the intersection of a set of inclusion-maximal subsets of the belief base H. The notion of a maximal consistent subset is an instance of an inclusion-maximal subset. Hence, sceptical reasoning with preferred subtheories requires us to solve a subproblem of type (i), while super-sceptical reasoning is based on solving a subproblem of type (ii).

Both subproblem (i) and subproblem (ii) are **coNP**-complete if we avail ourselves of an oracle for the consistency test. The consistency test is **NP**-complete for propositional logic, as explained above. Hence, standard belief base revision schemes are $\mathbf{coNP}^{\mathbf{NP}}$-complete when propositional logic is used to express beliefs. By a standard belief base revision scheme we mean one that requires solving the just explained subproblems. As the belief revision scheme of preferred subtheories is standard in this sense, belief revision with preferred subtheories is $\mathbf{coNP}^{\mathbf{NP}}$-complete as well. To be more precise:

Proposition 7.1 *If $K(\Theta)$ is defined by $\mid\!\sim^{PS(s)}$, the problem of deciding if $\phi \in K(\Theta) \oplus_i \alpha$ is $\mathbf{coNP}^{\mathbf{NP}}$-complete in propositional logic.*

Proposition 7.2 *If $K(\Theta)$ is defined by $\mid\!\sim^{PS(su)}$, the problem of deciding if $\phi \in K(\Theta) \oplus_i \alpha$ is $\mathbf{coNP}^{\mathbf{NP}}$-complete in propositional logic.*

Proofs using theorems in Nebel [89] are given in the Appendix.

Another noteworthy consequence of this analysis is that restricting the logic to propositional Horn logic does not suffice to achieve tractability. Only on condition of a linear ordering of the belief base, complexity reduces to **P** for propositional Horn logic. Such an ordering, however, seems hardly realistic since it would mean that there cannot be two beliefs with the same epistemic priority. More generally, we have already observed in the Introduction that all tractable belief revision schemes make either highly counter-intuitive assumptions about the epistemic ranking or have highly counter-intuitive outcomes (Sect. 1.7). Hence, we have concluded that there is no reasonably expressive tractable belief revision scheme to be found in the literature.

Relatively little research has been done on *resource-bounded belief revision*. This type of research takes cognitive limitations of real agents into account, but is often not taken to the development of a tractable belief revision scheme. Wassermann [115], for example, considers cognitive limitations of real agents, but gives no analysis of the computational complexity of revisions. The strategy is to perform contractions and revisions only in the context of a set of relevant beliefs. Alechina et al. [2] have devised a tractable account of belief revision in rule-based systems by means of recording justifications for beliefs. But this account is severely limited insofar as only literals can be retracted. Rules, by contrast, cannot be retracted. This implies that the account is not expressive enough to capture basic examples of default and nonmonotonic reasoning. For this to be seen, let us take a look at the famous Tweety example. Suppose we get to know that Tweety is a penguin, after we inferred her ability to fly from knowing that she is a bird. Then, according to the account by Alechina et al. [2], we have to retract that Tweety is a penguin or that Tweety is a bird, or both. None of these options accords with our actual pattern of default reasoning, which consists in retracting the application of the default rule. So, either we do not accept the default that all birds can fly, or we are forced to make counter-intuitive retractions. In any case, this is not how default reasoning works.

As is well known in the community of belief revision and nonmonotonic reasoning, we can define belief changes on the basis of certain reasonably expressive nonmonotonic logics [98, Chap. 3]. In light of the intractability of the more expressive belief revision schemes, it does not come as a surprise that reasonably expressive nonmonotonic logics are likewise intractable [28]. Note, in particular, that the sceptical inference relation of default logic remains intractable even on certain, relatively severe restrictions on the logical form of defaults [108]. This result is noteworthy because the sceptical inference relation of default logic is expressive enough to define belief changes in the spirit of AGM [7]. The work of Cadoli [27] on tractable approximations to determining the nonmonotonic inferences in default logic is confined to the credulous inference relation, which is too weak to define belief changes because it may generate inconsistent belief sets.

7.1.3 Cognitive Aspects

Why should the apparent intractability of belief revision be a concern from the perspective of philosophical logic? If philosophical logic aims at an account of the laws of truth, tractability would not be an issue at all. On this view, tractability and decidability are issues of applying logic, which can easily be separated from the pure core of logic. Famously, Frege [60] propounded the view that logic is the study of the laws of truth as opposed to the laws of accepting sentences. Hence, his anti-psychologism.

Cognitive aspects, however, appear immediately relevant if one views logic as the study of proper inferential reasoning. This view has paved the way for a number of very interesting developments in logic, such as formal systems for abductive, inductive, nonmonotonic reasoning and the whole field of dynamic epistemic logic to which belief revision is closely connected. In view of these developments, van Benthem [112, p. 67] has come to speak of a *cognitive turn* in logic, which is side-stepping Frege's anti-psychologism. This turn is driven by awareness of empirical facts and empirical problems in the broader field of human and automated reasoning, including computational issues. Take the example of nonmonotonic logics. Such logics are devised to model reasoning with incomplete information. If there was no such incompleteness, these logics would lose their point. Likewise, there is no motivation for studying belief revision for omniscient agents. Thus, the gap between the normative demands of classical logic and the limitations of our cognitive capacities has become a subject matter of research within philosophical logic to the effect of new formal systems that are presumed to be cognitively more appropriate than classical logic. Studying means to achieve tractability, we suggest, is a mere continuation of this line of research.

Why is complexity theory in computer science relevant for our understanding of human cognition? The connection between the two areas can be established in two ways. First, by the Church-Turing thesis. Second, by the computational theory of mind (CTM), according to which human cognition can be understood in terms of classical computations. The first connection is straightforward: the Church-Turing thesis says that any physically realisable computation device – whether it is based on silicon, neurones, DNA or some other alien technology – can be simulated by a Turing machine [14, p. 26]. In its strong form, the Church-Turing thesis says that this simulation takes only polynomial overhead. This strong version of the Church-Turing thesis implies that, if a problem is tractable for a non-Turing machine, it must be so for a Turing machine. By contraposition, this thesis implies that any problem intractable for Turing machines is intractable for the human mind as well.

The computational theory of mind gives a more vivid connection between complexity theory and human cognition. Fodor [58, p. 19] has characterised this theory by the following main principles:

(1) Thoughts have their causal roles in virtue of, inter alia, their logical form.
(2) The logical form of a thought supervenes on the syntactic form of the corresponding mental representation.

(3) Mental processes (including, paradigmatically, thinking) are computations, that is, they are operations defined on the syntax of mental representations, and they are reliably truth-preserving in indefinitely many ways.

Given CTM, we only need a minimum of naturalism to render the theory of computational complexity applicable to human cognition. Suppose the actual working of human minds is based on processes whose operations do not take infinitesimally small time. This is a very weak thesis, which is nowadays rarely contested and defeasibly confirmed by our biological theories. In conjunction with CTM, this thesis implies that computational complexity theory applies to human cognition.

In the Introduction, we have observed that the set of the following claims and theories is inconsistent: CTM, the theory of computational complexity (including the Church-Turing thesis in its strong form), $\mathbf{NP} \neq \mathbf{P}$, and the view that human minds are capable of obeying the norms of AGM-style belief revision theory and first-order logic (at least approximately and when confined to finite domains). Similar observations have been made by Cherniak [36], Fodor [58], Johnson-Laird [74], Stenning and van Lambalgen [107] and Woods [120]. According to Fodor [58, Chap. 2 and 3] and Fodor [59, Chap. 4.4], the computational infeasibility of belief revision is the central open problem of cognitive science, where cognitive science is defined by the view that human cognition is information processing.

As is obvious, there is a radical way of resolving the problem. One could simply deny that logic and belief revision theory have any role to play in an account of human cognition. We have argued against drawing such a radical consequence by pointing out that it is undeniable that both everyday and scientific reasoning exhibit an inferential structure. Accepting $\mathbf{NP} \neq \mathbf{P}$ and the other members of the above set of claims and theories, we have suggested working on the cognitive adequacy of logical systems and belief revision schemes themselves. Hence, the objective of this and the next chapter is twofold: first, devising a logic of modular reasoning, called *frame logic*. Second, devising a belief revision scheme that is tractable for frame logic, provided the syntactic size of the modular units of reasoning is bounded by a constant. Both frame logic and the tractable belief revision scheme are motivated by a truth maintenance system for the structuralist belief revision scheme to be devised in the remainder of this chapter. Working out such a truth maintenance system may seem more of an exercise in knowledge engineering, but this exercise will help us develop a novel logic eventually.

One clarification concerning reasoning about infinite domains is in order. We do not advocate a finitism that condemns all reasoning going beyond the limits of tractability and decidability. Finding a proof for a mathematical theorem will presumably never become a tractable problem in non-trivial domains of mathematics. Finding a unifying theory for related but different phenomena may never be handed over to an algorithm that has polynomial complexity. However, there are belief changes in everyday life and science that are performed fast and with a high degree of convergence, at least apparently so. Take the example of reading a newspaper.

The information update that is required for this better be tractable.[2] A scientist conducting an experiment is usually aware of the revisions that will follow from the possible outcomes. Also, her colleagues in the respective community are often in agreement about the changes that are necessary in light of new experimental results. Disagreement may arise from non-cognitive concerns and diverging preferences for certain paradigms. What we aim to explain, therefore, are only the ordinary convergent belief changes in normal science and everyday contexts, where all paradigms and priorities are consensually settled.

7.2 A Truth Maintenance System for Frames

7.2.1 Unique Model Extensions

To get started with devising a TMS using frames, let us make a preliminary assumption concerning the set of **T**-extensions of an intended application that are models of **T**. As we shall see, revisions can be performed much easier if it holds, for all intended applications y, that there is exactly one **T**-extension x of y that satisfies the substantial law of **T**. This property deserves explicit introduction:

Definition 7.1 (Unique model extension) We say that an intended application y of **T** has a unique model extension if and only if $EXT(\mathbf{T})(y) \cap \mathbf{M}(\mathbf{T})$ is a singleton.

By a model extension of an intended application we simply mean a **T**-extension that is a model of **T**. For the time being, all intended applications that do not have a unique model extension are ignored. At a later stage, we shall lift this assumption. Note that any instance of measuring a quantity by a particular method can be represented by an intended application that has the property of a unique model extension. For example, when **LP** (lever principle) is used for measuring the mass of a body, the model extension of this application will be uniquely determined.

7.2.2 Logical Foundations of the TMS

It goes without saying that it is the task of our TMS to compute belief changes on the basis of the structuralist belief revision scheme. That is, we assume that beliefs are represented by a structuralist default theory $\Theta(\mathbf{N})$. The formation and revision of beliefs is understood in the sense of Definition 6.4. As for the inferential closure of $\Theta(\mathbf{N})$, we chose the super-sceptical inference relation.

---

[2]Katsuno and Mendelzon [75] have argued for distinguishing between belief updates and belief revisions. Their argument, however, depends on not explicitly expressing the time for which a statement is held true. We assume, therefore, that belief revisions subsume updates.

Hence, the central problem to be solved by the TMS is to determine the set of successful intended applications. In other words, the TMS needs to determine which intended applications are a member of all preferred subtheories of a given structuralist default theory $\Theta(\mathbf{N})$. Only intended applications with this property contribute to the formation of derived beliefs in the super-sceptical inference relation. Such are the logical foundations upon which the following attempt at a truth maintenance system is based.

7.2.3 Determining Successful Intended Applications

Let us tackle the algorithmic determination of the set of successful intended applications. To this end, a few symbols need to be introduced. As in Chap. 6, $H(\mathbf{N})$ denotes the set that contains precisely the affirmed sentences of the form $b \in \mathbf{I}(\mathbf{T})$ for any $\mathbf{T}$ of $\mathbf{N}$. Further, let $H_s(\mathbf{N})$ be the set that "collects" the successful intended applications of $\mathbf{N}$. For sentences α of the form $b \in \mathbf{I}(\mathbf{T})$, $\mathbf{T}(\alpha)$ has the obvious meaning of denoting the theory-element that α refers to, while $\mathbf{I}(\alpha)$ denotes the intended application b itself. For simplicity, we call α an intended application when α is a sentence of the form $b \in \mathbf{I}(\mathbf{T})$. Strictly speaking, α is a sentential representation of an intended application in this case. This is a very preliminary algorithm that determines the set of successful intended applications:

Algorithm 7.1 (Determining successful intended applications)

```
H_s(N) := ∅;
for all α ∈ H(N):
  if α has a unique model extension:
    add α to H_s(N);
    for all ψ ∈ H_s(N) such that α ≠ ψ:
      for all links L:
        test (α, ψ) for satisfaction of link L;
        in case of non-satisfaction:
          if not α < ψ:            // if α is not epistemically superior to ψ
            delete α from H_s(N);
          if not ψ < α:
            delete ψ from H_s(N).
```

The basic idea of this algorithm is to add an intended application α to the set of successful intended applications just in case two conditions are satisfied: (i) α has a unique model extension, and (ii) all links between α and any application $\psi \in H_s(\mathbf{N})$ are satisfied, or the revision formula α has epistemic priority over ψ, or both. If a link is violated for a pair (α, ψ) and neither $\alpha < \psi$ nor $\psi < \alpha$ holds, both α and ψ are eliminated. This is an application of the famous WIDTIO principle: *when it doubt, throw it out.*[3]

[3]Winslett [118] introduces this phrase to characterise various kinds of intersection operations among potentially conflicting sentences and sets of sentences.

Note that quantifying phrases like *for all* $x \in A$: indicate that all subsequent, indented commands are to be executed for all $x \in A$. For this to be accomplished, some linear order of the members of A must be specified, but we shall not make this order explicit unless necessary.

$H_s(\mathbf{N})$ is an identifier of what computer scientists call a *dynamic set*, i.e., a set upon which such operations as deletion, addition and sorting of members can be performed. Both $H_s(\mathbf{N})$ and $H(\mathbf{N})$ figure as global variables, i.e., variables whose content can be accessed and modified in the whole program and not only in certain procedures. Using global variables dates back to non-object-oriented programming as it assumes a clear separation of data structures and methods. It greatly helps simplify the account of algorithms in our context, however. More use of global variables will be made in what follows.

All algorithms will be given in a semiformal fashion, in which informal phrases are used together with commands of familiar conventional programming languages. In other words, we write pseudocode. For the sake of brevity, colons and indenting are used to indicate blocks of commands in place of C's curly braces and Pascal's begin-end notation. The members of such a block are to be executed sequentially. Commands within a block are separated by semicolons. To maximise accessibility, use of standard commands of programming languages is kept to a minimum.

7.2.4 Revision by an Intended Application

The set of successful intended applications is a finite representation of beliefs in our system. We therefore need to determine how the set of successful intended applications changes upon new epistemic input. There is a straightforward way of doing this: simply add the revision formula α to the set $H(\mathbf{N})$ and determine anew the set $H_s(\mathbf{N})$ by the above algorithm. This strategy qualifies as a strict application of the direct mode of base revision. From a computational perspective, however, this strategy may not be optimal because it requires the generation of derived beliefs from scratch whenever new information comes in. Also, it is questionable whether this belief revision strategy is appropriate from a cognitive perspective. As a matter of fact, we often retain derived beliefs without having to generate them anew from our basic ones in the context of background theories. We must therefore wonder how we can determine which of the derived beliefs are to be retracted in the course of a revision. With only minor modifications, we can turn Algorithm 7.1 into an algorithm for the revision of $H_s(\mathbf{N})$:

Algorithm 7.2 (Revision of $H_s(\mathbf{N})$ by α)

```
revise(α)
if α has a unique model extension:
  add α to H_s(N);
  for all ψ ∈ H_s(N) such that ψ ≠ α:
    for all links L:
```

test (α, ψ) for satisfaction of link **L**;
in case of non-satisfaction:
  if not $\alpha < \psi$:
    delete α from $H_s(\mathbf{N})$;
    break; // leave the present loop
    break;
  if not $\psi < \alpha$:
    delete ψ from $H_s(\mathbf{N})$;
    break.

This algorithm makes essential use of the presently accepted division into successful and unsuccessful intended applications as well as predetermined links among intended applications. Thereby it circumvents determining the set $H_s(\mathbf{N})$ from scratch.

The command *break* has the effect of leaving the innermost loop that is presently being executed. Once we realise that α is not consistent with some successful intended application that is superior to α, we must consider α unsuccessful. Hence, we can leave both loops within which the test of link satisfaction is performed for α.

7.2.5 Link Satisfaction

To properly explain the meaning of all elements of the TMS, let us recall what it means for some link between two intended applications to be satisfied. We have come to confine our understanding of links to equality links. Hence, satisfaction of all links for a pair (α, β) of intended applications means that there is a pair (x_α, x_β) of admissible **T**-extensions such that x_α is semantically consistent with x_β. To be more precise:

Explanation 7.1 (Link satisfaction between applications) Let α and β be two applications, and $\mathbf{A}(\alpha)$ and $\mathbf{A}(\beta)$ the respective sets of admissible extensions. All links between α and β are satisfied if and only if there are $x_\alpha \in \mathbf{A}(\alpha)$ and $x_\beta \in \mathbf{A}(\beta)$ such that there are no tuple $(a_1, \ldots, a_n)$ and no relation concept R such that

(1) $(a_1, \ldots, a_n) \in (x_\alpha)_R$
(2) $(a_1, \ldots, a_n) \notin (x_\beta)_R$
(3) there is $x \in EXT(\mathbf{T}(\beta))(\mathbf{I}(\beta))$ such that $(a_1, \ldots, a_n) \in (x)_R$

or

(1) $(a_1, \ldots, a_n) \in (x_\beta)_R$
(2) $(a_1, \ldots, a_n) \notin (x_\alpha)_R$
(3) there is $x \in EXT(\mathbf{T}(\alpha))(\mathbf{I}(\alpha))$ such that $(a_1, \ldots, a_n) \in (x)_R$.

Semantic consistency of two structures x_α and x_β, thus, means that, for all relation concepts R, the interpretation of R by x_α is semantically consistent with the interpretation of R by x_β. Semantic consistency can be tested in a straightforward manner if the sets $(x_\alpha)_R$, $(x_\beta)_R$, $EXT(\mathbf{T}(\alpha))(\mathbf{I}(\alpha))$ and $EXT(\mathbf{T}(\beta))(\mathbf{I}(\beta))$ are finite. And the number of computation steps needed for this test is bounded by a constant if the cardinality of these sets is bounded by a constant. $(x_\alpha)_R$ ($(x_\beta)_R$) stands for the interpretation of the relation concept R by the structure x_α (x_β). Henceforth, for all intended applications α, let $\mathbf{A}(\alpha)$ designate the set of model extensions that are considered admissible at the respective stage of computation.

Specialisation links would not be difficult to take into account but are disregarded in the present exposition of the TMS for the sake of simplicity. We shall assume, moreover, that there are no plain contradictions among intended applications in the sense that two intended applications (conceived as partial structures) are not semantically consistent with one another. On this condition, failure of link satisfaction can only arise from applying substantial laws of theory-elements to intended applications. We could lift these two assumptions at the price of a few more tests, but these would not give us much insight.

7.2.6 Refinements

Algorithm 7.2 is still but a first approximate solution to the problem of determining and revising the set of successful intended applications. A proper solution ought to take a number of further aspects into account. First, one must not ignore, as the algorithm does, the possibility of an unsuccessful intended application α_1 that was blocked by a successful intended application α_2 that all of a sudden gets blocked by some α_3. (We say that α_2 blocks α_1 if and only if the pair (α_1, α_2) fails to pass the test of link satisfaction, and α_1 has a lower epistemic rank than α_2.) In this case, α_1 may become a successful intended application through the retraction of α_2. For example, once we are convinced that a certain measurement m is faulty, we may have to reconsider some hypothesis h that was ruled out just because of m. That is, we should revise $H_s(\mathbf{N})$ by the intended application that represents h in order to see whether it may succeed, then.

Second, we may have intended applications whose relations are, in part, determined by links to other intended applications. More precisely, it is the admissible $\mathbf{T}$-extensions of an intended application α_1 that may determine the relations of another intended application α_2. In classical structuralism, one speaks of links determining the $\mathbf{T}$-non-theoretical relations of intended applications in such a case. Hence, a more refined variant of the algorithm should allow for the determination of intended applications through links.

Third, assume α is a successful intended application that determines some relation of some other intended application β. Then, in case α becomes blocked in the course of a revision, the determination of β by α needs to be withdrawn. Once we become convinced that a certain measurement or hypothesis is faulty (because it has been ruled out by reliable evidence), we should retract those determinations that were based on this measurement or hypothesis. This problem is not trivial at all since it seems as if we need to keep track of all determinations of any given intended application in order to retract such determinations if necessary.

Fourth, let us take a closer look at the manner in which the algorithm makes use of the WIDTIO principle. Assume $H_s(\mathbf{N})$ is revised by α_1, α_2 and α_3 sequentially, which all have equal reliability. Further, α_1, α_2 and α_3 may be mutually inconsistent with one another but consistent with any member of $H_s(\mathbf{N})$. Now, there are six different total orders of these intended applications, each of which represent a possible temporal order of corresponding revision operations: $\langle\alpha_1, \alpha_2, \alpha_3\rangle$, $\langle\alpha_1, \alpha_3, \alpha_2\rangle$, $\langle\alpha_2, \alpha_1, \alpha_3\rangle, \ldots, \langle\alpha_3, \alpha_2, \alpha_1\rangle$. The problem that we see surfacing concerns the WIDTIO principle, according to which deletions of members of $H_s(\mathbf{N})$ are performed in Algorithm 7.2: the two intended applications that come in first will be expelled from $H_s(\mathbf{N})$, while the third one is accepted as successful. Such a dependency of beliefs on the temporal order of revisions does not seem acceptable. This will become particularly obvious in light of examples that we shall discuss below.

In sum, we need to complement the present algorithm with at least four further operations. First, an operation that reconsiders those unsuccessful intended applications that have been blocked by an intended application that happens to be retracted from the set of successful ones at a later stage. Second, the operation of determining intended applications by means of links. Third, the reverse operation that retracts such determinations in case the determining intended application becomes rejected. Finally, it is desirable to have a refined version of the deletion of intended applications from $H_s(\mathbf{N})$ such that the determination of the set of successful intended applications does not depend on the temporal order of revisions.

7.2.6.1 Recursive Revisions

Let us get to work on the first of the refinements just explained. In case an intended application β is retracted, we need to test whether it is linked with another intended application ψ that has been blocked by β because of a failure of link satisfaction. The relation of blocking here means that ψ was retracted from $H_s(\mathbf{N})$ because there is a link $\mathbf{L}$ such that (β, ψ) does not satisfy $\mathbf{L}$ and β is epistemically superior to ψ. Now, ψ may succeed once again when β is retracted from $H_s(\mathbf{N})$. Let $H_u(\mathbf{N})$ be the dynamic set that collects the unsuccessful intended applications. Further, let $B(\beta)$ designate the set that collects, for any intended application β, the intended applications that have been blocked by β. We write $\beta < \psi$ for β being epistemically superior to ψ since we understand the epistemic ranking of a default theory $\Theta = \langle T_1, \ldots, T_n\rangle$ such that the members of T_1 have the highest epistemic rank. We introduce the new procedure **reject** for the refinement to work:

Algorithm 7.3 (Recursive revision of $H_s(\mathbf{N})$ by α)
  revise(α)
  if α has a unique model extension:
    add α to $H_s(\mathbf{N})$;
    for all $\psi \in H_s(\mathbf{N})$ such that $\psi \neq \alpha$:

```
for all links L:
  test (α, ψ) for satisfaction of link L;
  in case of non-satisfaction:
    if not α < ψ:                // that is, α is not epistemically superior to ψ
      reject(α, ψ);
      break;
      break;
    if not ψ < α:
      reject(ψ, α)
      break.
```

```
reject(δ, ε)                     // δ is rejected because of an inconsistency with ε
add δ to B(ε);
delete δ from H_s(N);
add δ to H_u(N);
for all ψ ∈ B(δ) such that ψ ≠ ε:
  revise(ψ);
B(δ) := ∅.
```

Note that this revision algorithm has a recursive structure: while executing **reject**, which is a subprocedure of **revise**, **revise** may be called. By means of this recursive structure, it is possible to test whether the retraction of an intended application from the set $H_s(\mathbf{N})$ unblocks another intended application. We say that an intended application ψ becomes unblocked if it was rejected by some β, which is then retracted.

7.2.6.2 Determination Through Links

There are intended applications that have no unique model extension but may be determinable by links to the effect that the resulting intended application has this property. To give an example from the domain of metals and acids, we may observe that some golden piece of metal becomes stained through exposition to some liquid of unknown chemical composition. Later on, we come to conclude that the staining liquid was sulphuric acid as we observe a white precipitation in the presence of Ca^{2+} ions. This is a standard test procedure for H_2SO_4. It is not until the second empirical observation that we have to conclude that the golden piece of metal is not made of gold.

To indicate how this example translates into our system, let $\mathbf{T}_1$ encode the law that gold is not stained by sulphuric acid and $\mathbf{T}_2$ the law that SO_4^{2-} ions can be precipitated by means of Ca^{2+} ions. Further, let α_1 and α_2 designate the intended applications under consideration, where $\mathbf{T}(\alpha_1) = \mathbf{T}_1$ and $\mathbf{T}(\alpha_2) = \mathbf{T}_2$. Then, we can say that the intended application α_1 is determined through a link to α_2 since it is only by α_2 that we know the type of substance that stained the piece of metal. Likewise, in classical structuralism, one envisions to have intended applications whose **T**-non-theoretical relations are determined by links. In simple cases, determination of intended applications through links consists in a transfer of data from one intended application to another.

Determination of intended applications through links arises from the requirement of link satisfaction. Assume it holds for all intended applications that the number of model extensions is finite. Consider now the test of link satisfaction for two intended applications α and β. Any $x_1 \in \mathbf{A}(\alpha)$ for which there is no $x_2 \in \mathbf{A}(\beta)$ such that (x_1, x_2) satisfies all links is not admissible. Likewise, any $x_1 \in \mathbf{A}(\beta)$ for which there is no $x_2 \in \mathbf{A}(\alpha)$ such that (x_1, x_2) satisfy all links, is not admissible either. These two claims about the admissibility of extensions follow from the postulate schema $\mathbf{P4}(\mathbf{L}_i, \mathbf{T}_j, \mathbf{T}_k)$:

$$\forall x_1 \forall y_1 \forall x_2 \forall y_2 ((x_1, y_1) \in \mathbf{AE}(\mathbf{T}_j) \wedge (x_2, y_2) \in \mathbf{AE}(\mathbf{T}_k) \rightarrow \phi(\mathbf{L}_i)(x_1, x_2))$$

where $\phi(\mathbf{L}_i)(x_1, x_2)$, if it is an equality link, says that the structures x_1 and x_2 are consistent with one another concerning the valuation of a certain relation concept (cf. Sect. 7.2.4). Hence, given two intended applications α and β, a $\mathbf{T}$-extension x_1 of α is only admissible if there is an admissible $\mathbf{T}'$-extension x_2 of β such that all links are satisfied for (x_1, x_2), and vice versa. The determination of $\mathbf{A}(\alpha)$ thus aims at determining, for all intended applications α, the set of admissible $\mathbf{T}$-extensions of α.

In Sect. 7.2.7, we shall specify a procedure **satisfaction** that has two functions. First, it tests satisfaction of all links for a given pair (α, β) of intended applications. Second, in doing so, structures $x_1 \in \mathbf{A}(\alpha)$ $(\mathbf{A}(\beta))$ that are not consistent with any structure $x_2 \in \mathbf{A}(\beta)$ $(\mathbf{A}(\alpha))$ are deleted from the set $\mathbf{A}(\alpha)$ $(\mathbf{B}(\beta))$, i.e., the set of admissible extensions of α (β). So, presumably admissible $\mathbf{T}$-extensions of an intended application α that violate some link to all presumably admissible $\mathbf{T}'$-extensions of another intended application β must be deleted from the set of admissible extensions. In the course of such deletions, intended applications become more and more determined. *Omnes determinatio est negatio*, as Spinoza has taught us. Henceforth we assume that, for all intended applications, the set of model extensions is finite.

7.2.6.3 Retracting Determinations

Suppose some intended application α_1 determines some intended application α_2. After a few revisions, α_1 becomes retracted because it does not satisfy all links to some intended application α_3 being superior to it. Then, the determination of α_2 by α_1 needs to be retracted as well. To take such retractions into account, two further types of dynamic sets need to be introduced. First, a set $DI_a(\phi)$ that "collects" the intended applications that are determined by ϕ, for all intended applications ϕ. Second, a set $DI_p(\psi)$ may collect all intended applications trough which ψ is determined, for all intended applications ψ. So, subscript a stands for active determination and p for passive determination. By convention, for all intended applications ϕ, $\phi \in DI_p(\phi)$ and $\phi \in DI_a(\phi)$.

An algorithm that retracts determinations may work as follows:

(1) If α becomes retracted, all members of $DI_a(\alpha)$ need to be redetermined.
(2) Redetermination of some $\phi \in DI_a(\alpha)$ consists of three steps:

   (a) α is deleted from $DI_p(\phi)$.
   (b) $\mathbf{A}(\phi)$ is set to the value prior to any determination through links.
   (c) **satisfaction** must be called for all pairs (ϕ, ψ) with $\psi \in DI_p(\phi)$.
   (d) In case the original value of $\mathbf{A}(\phi)$ cannot be recovered insofar as it has been changed through the last two steps: all direct and indirect determinations of ϕ need to be retracted.

7.2.6.4 Updating the Epistemic Ranking

When determination of intended applications through links is allowed, it becomes necessary to update the epistemic rank of intended applications thus determined. Suppose we have an intended application α_1 of a theory-element $\mathbf{T}_1$ with a very high degree of reliability. Then, α_1 may be determined by some intended application α_2 whose theory-element $\mathbf{T}_2$ has a lower degree of reliability. Once α_1 is determined by α_2, a conflict arises between α_1 and another successful intended application α_3.

What shall we take as the epistemic rank of α_1? That of $\mathbf{T}_1$ or that of $\mathbf{T}_2$ or some combination of the two? Since α_1 is determined, in part, by α_2 and since the theory-element of α_2 is less reliable than that of α_1, we should not take the epistemic rank of $\mathbf{T}_1$ as the degree of reliability of α_1. Rather, it is reasonable to take the minimum of the ranks of $\mathbf{T}_1$ and $\mathbf{T}_2$ as the degree of reliability of α_1. And if there is a conflict between α_1 and α_3 such that α_1 is less reliable than α_3 (since α_2 has lower epistemic rank than α_3), it is α_2 in place of the stronger α_1 that needs to be retracted from the set of successful intended applications. This follows from the principle that an argument is just as strong as its weakest premise is.

To resume our example from the domain of acids, suppose the test that was performed to identify the chemical substance of the staining liquid is considered to have low reliability. Arguably, precipitation of SO_4^{2-} ions does not rule out that the staining liquid consists of aqua regia to a very high percentage. The claim that the ring is made of gold, however, may not be obtained by a report of a jeweller in Casa Blanca. It is rather based on, first, measuring the density of the ring and, second, verifying that the ring has no hidden core of a silvery metal by heating it beyond the melting point of gold. This is a fairly reliable test on condition that the values of mass and volume are determined with a very high degree of precision. For, there are very few chemical elements that have a higher density than gold. These elements (rhenium, osmium, iridium, platinum, neptunium and plutonium) are very rare and do not look golden. In this case, we should retract the chemical identification of the staining liquid but not the proposition that the ring is made of gold. This corresponds to a retraction of α_2 in the above scenario.

What are we to do if α_2 and α_3 have the same degree of reliability in the above scenario? Clearly, we should doubt both α_2 and α_3. Hence, α_2 and α_3 would have to be retracted from $H_s(\mathbf{N})$. Let us cast these considerations into general principles:

(1) We say that an intended application α_1 determines directly another intended application α_2 if and only if the test of link satisfaction for (α_1, α_2) leads to diminishing the set $\mathbf{A}(\alpha_2)$. α_1 determines α_n indirectly if and only if there are $\alpha_1, \ldots, \alpha_n$ $(n > 2)$ such that α_i determines α_{i+1} directly for all i $(1 \leq i < n)$.
(2) Let $D(\alpha)$ be the set of intended applications that determine, directly or indirectly, some intended application α. For all intended applications α, $\alpha \in D(\alpha)$. That is, every intended application is determined by itself. Let $DI_w(\alpha)$ designate the set whose members have minimal reliability among the members of $D(\alpha)$. So, $DI_w(\alpha) := \{\phi \in D(\alpha) \mid$ there is no $\psi \in D(\alpha)$ such that $\mathbf{T}(\psi) < \mathbf{T}(\phi)\}$.
(3) Let $D(\alpha)$, $DI_w(\alpha)$ and α be as just introduced. Further, we may have some intended application β with corresponding sets $D(\beta)$ and $DI_w(\beta)$. In case α is in conflict with β, all members of $DI_w(\alpha)$ need to be retracted from $H_s(\mathbf{N})$ if there are $\psi_1 \in DI_w(\alpha)$, $\psi_2 \in DI_w(\beta)$ such that $\psi_2 < \psi_1$.
(4) Let α and β be as just explained. If they are in conflict with one another, and there are no $\psi_1 \in DI_w(\alpha)$, $\psi_2 \in DI_w(\beta)$ such that either $\mathbf{T}(\psi_2) < \mathbf{T}(\psi_1)$ or $\mathbf{T}(\psi_1) < \mathbf{T}(\psi_2)$, all members of both $DI_w(\alpha)$ and $DI_w(\beta)$ are to be retracted.

These principles may be translated into the following additional operations of our TMS:

(1) Whenever some intended application α_1 is determined by some intended application α_2, a procedure **weakest** is to be called that determines the weakest elements of $DI_w(\alpha_1) \cup DI_w(\alpha_2)$. These elements form the new set $DI_w(\alpha_1)$.
(2) When some α_2 is inconsistent with some α_3, the epistemic ranks of the members of $DI_w(\alpha_2)$ and $DI_w(\alpha_3)$ need to be compared in order to decide which intended application, or which intended applications, are to be rejected.
(3) When an intended application α that determines ϕ is retracted, $DI_w(\phi)$ needs to be updated.

Before we spell out these additions, let us take a closer look at the WIDTIO principle.

7.2.6.5 Refining WIDTIO

The present use of WIDTIO was found to imply a counter-intuitive dependency of beliefs on the temporal order of incoming intended applications. More precisely, it was assumed that three intended applications of equal reliability are mutually incompatible with one another, while being consistent with the current set of successful intended applications. Then, the two intended applications that come first will not remain in the set $H_s(\mathbf{N})$, while the last one does. The problem recurs with more than three mutually inconsistent intended applications, of course. The solution we propose is to distinguish between doubting and rejecting an intended application:

(1) An intended application α becomes *rejected* (in the sense of being considered unsuccessful and thus becoming a member of $H_u(\mathbf{N})$) just in case there is some successful intended application that is superior to α and inconsistent with α.
(2) An intended application α is *doubted*, by contrast, just in case it is not rejected but inconsistent with some unrejected intended application β to which α is not superior.

Rejected intended applications can be ignored when determining the success of a new incoming intended application. We require, however, that a new incoming intended application α be only accepted as successful if it holds, for all doubted intended applications β, that α is consistent with or superior to β. In case α is consistent with all members of $H_s(\mathbf{N})$ but inconsistent with some doubted intended application β and $\alpha =_e \beta$, α must be doubted as well. ($\alpha =_e \beta$ means that α and β have the same epistemic rank.) Let $H_d(\mathbf{N})$ be the set of doubted intended applications.

We thus require, for a new intended application to be successful, consistency with or superiority to any member of both $H_s(\mathbf{N})$ and $H_d(\mathbf{N})$. Still, there remains a difference between successful and doubted intended applications. Only the former but not the latter contribute to the formation of derived beliefs. This is the key to resolving the above issue: $\alpha_1, \alpha_2, \alpha_3$ will all be doubted but not rejected, independently of their temporal order as revisions, given that each is consistent with any successful intended application and they are inconsistent with one another. They thus become members of $H_d(\mathbf{N})$.

7.2.6.6 Prioritising Revise

Recall that the construction of preferred subtheories begins with maximal consistent subsets of the most reliable beliefs, then moves on to add second most reliable beliefs in a consistent manner, and so on. We can respect this order of construction by prioritising the execution of the operation **revise**(α). That is, we first "collect" intended applications for which **revise** needs to be called, and then call **revise**, starting with the applications of the highest epistemic rank. For this to be achieved, a dynamic and totally ordered set $TR^{<}(\mathbf{N})$ is defined that collects intended applications for which **revise** needs to be called. TR stands for *to revise*.

Likewise, we can enhance the efficiency of the TMS by prioritising the test of link satisfaction. That is, the TMS becomes more efficient if we test the satisfaction of links between the revision formula α and applications $\psi \in H_s(\mathbf{N})$ in such a manner that we begin with those applications of $H_s(\mathbf{N})$ that have the highest epistemic priority, then move on to those with second highest priority, and so on. Thereby, we can avoid that α determines some ψ_1 of lower epistemic priority, but is rejected at a later stage by some ψ_2. In this case, the determination of ψ_1 by α would have to be retracted. Hence, it would have been more efficient not to let α determine ψ_1

in the first place. Therefore, testing link satisfaction should be prioritised in such a manner that the test starts with the applications of the highest epistemic rank. Similar considerations apply to the test of link satisfaction with the doubted applications.

To prioritise the test of link satisfaction, the TMS should keep track of the epistemic rank of the members of $H_s(\mathbf{N})$ and $H_d(\mathbf{N})$. Let, therefore, $H_s^<(\mathbf{N})$ and $H_d^<(\mathbf{N})$ designate the totally ordered, dynamic sets of successful and doubted applications, respectively. The total order of $H_s^<(\mathbf{N})$ and $H_d^<(\mathbf{N})$ must respect the epistemic ranking of their members. We continue to use $H_s(\mathbf{N})$ and $H_d(\mathbf{N})$ for the corresponding non-prioritised sets.

7.2.7 The Truth Maintenance Algorithm

The following algorithm takes care of (i) the retraction of determinations, (ii) updating the epistemic rank of an intended application, (iii) the distinction between rejecting and doubting an intended application, and (iv) the prioritised execution of **revise**.

Algorithm 7.4 (Refined recursive revision of $H_s^<(\mathbf{N})$ by α)

revise_in(α) // non-recursive initial procedure
$\mathbf{A}(\alpha) := EXT(\mathbf{T}(\alpha))(\alpha) \cap \mathbf{M}(\mathbf{T}(\alpha))$; // that starts a revision by α
$B(\alpha) := \emptyset$;
$DI_a(\alpha) := \{\alpha\}$;
$DI_p(\alpha) := \{\alpha\}$;
$DI_w(\alpha) := \{\alpha\}$;
if $\mathbf{A}(\alpha) \neq \emptyset$:
  insert α into $TR^<(\mathbf{N})$; // $TR^<(\mathbf{N})$ is empty when the revision starts
else:
  add α to $H_u(\mathbf{N})$;
while $TR^<(\mathbf{N})$ is not empty:
  call **revise** for the top element of $TR^<(\mathbf{N})$.

revise(α) // recursive revision procedure
$\mathbf{A}_2(\alpha) := \mathbf{A}(\alpha)$; // unlike $\mathbf{A}(\alpha)$, $\mathbf{A}_2(\alpha)$ is a local variable
insert α into $H_s^<(\mathbf{N})$;
for all $\psi \in H_s^<(\mathbf{N})$ such that $\psi \neq \alpha$ (in the order of
their epistemic ranking, starting with the highest):
  $\mathbf{A}_2(\psi) := \mathbf{A}(\psi)$; // unlike $\mathbf{A}(\psi)$, $\mathbf{A}_2(\psi)$ is a local variable
  sat:= false; // consistency flag for (α, β)
  satisfaction(α, ψ, sat);
  if not sat:
    compare(α, ψ);
  else:
    if $\mathbf{A}(\psi) \neq \mathbf{A}_2(\psi)$: // if $\mathbf{A}(\psi)$ has changed by calling **satisfaction**

```
        insert ψ into TR<(N);
      if A(α) ≠ A2(α):
        insert α into TR<(N);
        return;                                   // leave procedure revise
    if α ∈ Hu(N):
      return;

  if α ∈ Hs<(N):              // test of consistency with members of Hd<(N) starts
                                  // if α has not been rejected or doubted so far
    for all ψ ∈ Hd<(N) (in the order of their
    epistemic ranking, starting with the highest):
      A2(ψ) := A(ψ);
      sat:= false;
      satisfaction(α, ψ, sat);
      if not sat:
        compare(α, ψ);
      else
        if A(ψ) ≠ A2(ψ):
          insert ψ into TR<(N);
      if α ∈ Hu(N):
        break.


  satisfaction(α, β, sat)           // test of link satisfaction between α and β
  A'(α) := ∅;              // A'(α) is a local variable collecting all x1 ∈ A(α) for
                     // which an x2 ∈ A(β) is found s.t. (x1, x2) satisfies all links
  A'(β) := ∅;              // A'(β) is a local variable collecting all x2 ∈ A(β) for
                     // which an x1 ∈ A(α) is found s.t. (x1, x2) satisfies all links
  for all x1 ∈ A(α):
    for all x2 ∈ A(β):
      for all links L:
        if (x1, x2) does not satisfy L:
          break;
          break;
      add x1 to A'(α);
      add x2 to A'(β);

  if A'(β) ≠ ∅ and A'(β) ≠ ∅:
    sat:=true;
    if β ∈ Hs<(N) and A(α) ≠ A'(α):                  // β determines α only
      A(α) := A'(α);                                  // if β is successful
      add α to DIa(β);
      add β to DIp(α);
      weakest(β, α);
    if α ∈ Hs<(N) and A(β) ≠ A'(β):
```

```
    A(β) := A′(β);
    add β to DI_a(α);
    add α to DI_p(β);
    weakest(α, β).

weakest(φ, α)               // updating the set of weakest intended applications
                            // that determine α, after a determination of α by φ
take some φ_i ∈ DI_w(φ):               // since all members of DI_w(φ) have the
  take some α_j ∈ DI_w(α):             // same epistemic rank, it suffices
    if α_j < φ_i:                      // to compare some member of DI_w(φ)
      DI_w(α) := DI_w(φ);              // with some member of DI_w(α)
    if φ_i =_e α_j:                    // =_e stands for equal epistemic rank
      DI_w(α) := DI_w(α) ∪ DI_w(φ).

compare(α, ψ)      //called in case of a conflict between the revision formula α
                   // and some successful or doubted intended application ψ
take some α_i ∈ DI_w(α):
  take some ψ_j ∈ DI_w(ψ):
    if α ∈ H_s^<(N):
      if α_i < ψ_j:
        reject(ψ, α);                  // ψ is rejected by α
        return;                        // leave the presently executed procedure
      if α_i =_e ψ_j:
        doubt(α, ψ);                   // α is doubted by ψ
        if ψ ∈ H_s^<(N):
          doubt(ψ, α);
          return;
        else:
          return;
      if ψ ∈ H_d^<(N) and ψ_j < α_i:
        doubt(α, ψ);
        return;
    if ψ ∈ H_s^<(N):                   // this part is executed only if α ∈ H_d^<(N)
      if ψ_j < α_i:
        reject(α, ψ);
        return;
      if α_i < ψ_j:
        doubt(ψ, α).

reject(δ, ε)                 // δ is rejected because of an inconsistency with ε
for all δ_i ∈ DI_w(δ):
  for all γ ∈ DI_p(ε) ∪ DI_p(δ) \ DI_w(δ):
    add δ_i to B(γ);                   // γ is noted to block δ_i
```

  delete δ_i from $H_s^<(\mathbf{N})$; // one of these two operations
  delete δ_i from $H_d^<(\mathbf{N})$; // will be vacuously performed
  add δ_i to $H_u(\mathbf{N})$;
  for all $\psi \in B(\delta_i)$: // see if applications are unblocked
    reconsider(ψ, δ_i); // through undetermining α
undetermine_s($DI_w(\delta)$).

doubt(δ, ε) // δ is doubted because of an inconsistency with ε
for all $\delta_i \in DI_w(\delta)$:
  for all $\gamma \in DI_p(\delta) \cup DI_p(\epsilon) \setminus DI_w(\delta)$:
    add δ_i to $B(\gamma)$;
  delete δ_i from $H_s^<(\mathbf{N})$;
  insert δ_i into $H_d^<(\mathbf{N})$;
  for all $\psi \in B(\delta_i)$: // see if applications are unblocked
    reconsider(ψ, δ_i); // through doubting α
undetermine_s($DI_w(\delta) \cup DI_w(\varepsilon)$).

undetermine_s(Γ) // retracting the determinations of all members of Γ
for all $\gamma \in \Gamma$:
  undetermine(γ, Γ).

undetermine(γ, Γ) // retracting the determinations of γ
for all $\alpha \in DI_a(\gamma) \setminus \{\gamma\}$:
  delete α from $DI_a(\gamma)$;
  delete γ from $DI_p(\alpha)$;
  delete γ from $DI_w(\alpha)$;
  $\mathbf{A}_2(\alpha) := \mathbf{A}(\alpha)$;
  $\mathbf{A}(\alpha) := EXT(\mathbf{T}(\alpha))(\alpha) \cap \mathbf{M}(\mathbf{T}(\alpha))$; // set of admissible extensions of α
  // is set to its original value
  for all $\psi \in DI_p(\alpha) \setminus \Gamma$: // α is redetermined
    satisfaction(ψ, α, false);
  if $\mathbf{A}_2(\alpha) \neq \mathbf{A}(\alpha)$:
    for all $\psi \in B(\alpha)$:
      if $\psi \notin \Gamma$: // test if applications are unblocked
        reconsider(ψ, α); // through undetermining α
undetermine($\alpha, \emptyset$).

reconsider(ψ, α) // test if classification of ψ into $H_u(\mathbf{N})$, $H_d^<(\mathbf{N})$
delete ψ from $B(\alpha)$; // and $H_s^<(\mathbf{N})$ changes, now that α has been
delete ψ from $H_u(\mathbf{N})$; // undetermined, doubted, or rejected
delete ψ from $H_d^<(\mathbf{N})$;
insert ψ into $TR^<(\mathbf{N})$.

7.3 Tractability

Now that the details of the TMS have been worked out, we investigate its general properties. In this section, we show that the TMS determines belief changes in a tractable manner. Then, we move on to investigating whether this tractable computation of belief changes is sound and complete with respect to the structuralist revision system upon which the TMS is founded.

7.3.1 *Operation Tree*

The general operation of truth maintenance consists, in our system, of operations that are performed on intended applications. Upon a single call of **revise** for some intended application α, three kinds of operations may be carried out:

(1) determination of an intended application
(2) doubt of an intended application
(3) rejection of an intended application.

Let α be the revision formula. Then, α determines β if and only if (i) the pair (α, β) passes the test for link satisfaction, (ii) $\alpha \in H_s(\mathbf{N})$ at the time of this test, but (iii) there is $x_1 \in \mathbf{A}(\beta)$ that has no corresponding $x_2 \in \mathbf{A}(\alpha)$ such that (x_1, x_2) satisfies all links. Likewise for the case where some $\beta \in H_s(\mathbf{N})$ determines α. α doubts $\beta \in H_s(\mathbf{N})$ and vice versa if (i) (α, β) fails to pass the test for link satisfaction, (ii) $\alpha \in H_s(\mathbf{N})$ at the time of this test, and (iii) α and β have the same epistemic rank. Moreover, $\beta \in H_d(\mathbf{N})$ doubts $\alpha \in H_s(\mathbf{N})$ if (α, β) fails to pass the test for link satisfaction, and β has at least the epistemic rank of α. Finally, $\alpha \in H_s(\mathbf{N})$ rejects $\beta \in H_s(\mathbf{N}) \cup H_d(\mathbf{N})$ if and only if (α, β) fail to pass the test for link satisfaction, and β has a lower epistemic rank than α. Likewise for the inverse case where the revision formula α is rejected by some $\beta \in H_s(\mathbf{N})$.

Determination, in turn, results in the call of **revise** for the intended application determined. Doubting some β_2 will induce undetermination (i. e., the retraction of determinations) of all intended applications that have been determined by β_2, if there are any. Undetermination, in turn, requires the call of **revise** for those applications that were blocked by an application whose determination by β_2 is retracted. Also, **revise** must be called for all intended applications that have been blocked by β_2 so far. Likewise, rejecting some intended application β_3 induces undetermination of those intended applications that have been determined by β_3 and revision by all intended applications that have been blocked by β_3. Both undetermination and unblocking are recursive operations. With this in mind, we can represent the formal structure of the operations of the TMS by means of the tree of Fig. 7.1.

Edges represent operations that are performed upon an intended application and the dynamic sets $H_s(\mathbf{N})$, $H_u(\mathbf{N})$ and $H_d(\mathbf{N})$. *rev* abbreviates revision, *det* determination, *doubt* the operation of doubting, *rej* rejection and *undet* the operation of undetermination. The intended application taken by the operation as an argument

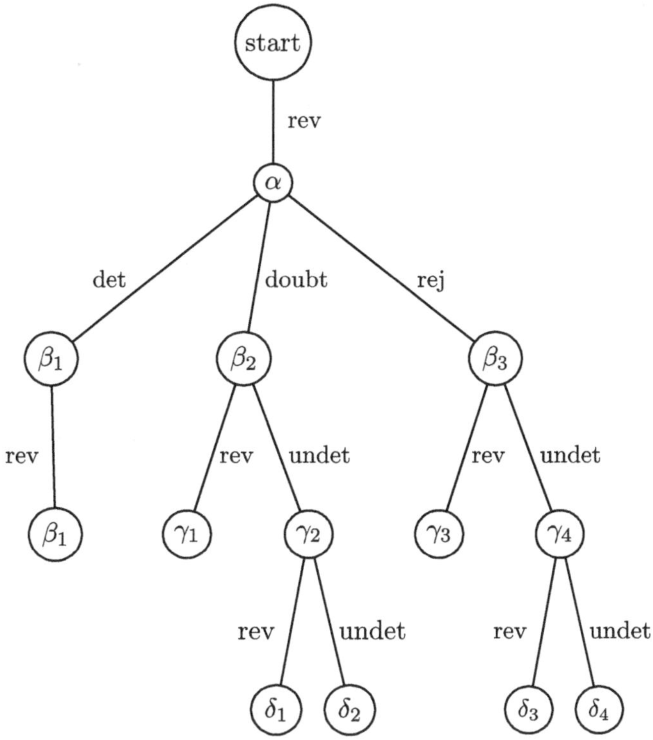

Fig. 7.1 Structure of the operation tree of the TMS

is represented by the edge's child node. The tree represents the formal structure of the TMS insofar as it indicates which types of operations may in principle be performed. Note that a single call of **revise** may induce several determinations, doubts and rejections. For, if α remains successful, all intended applications that are both inconsistent with α and inferior to it, must be rejected. Likewise, the rejection of some intended application may induce several undeterminations and revisions. Note, furthermore, that the revision by α may well result in determining α, doubting α and rejecting α. With respect to Fig. 7.1, this means that some of the applications β_1, β_2 and β_3 may be identical with α.

The working of the TMS may now be described as follows. A revision by α starts with testing the consistency of pairs (α, β) for all β in $H_s(\mathbf{N}) \cup H_d(\mathbf{N})$. Depending on the result of this test, the TMS induces operations of determination, doubt and rejection. Whenever such an operation changes some intended application β – in the sense of changing the dynamic set $\mathbf{A}(\beta)$ or changing membership of β as regards the sets $H_s(\mathbf{N})$ or $H_d(\mathbf{N})$ –, it must be checked whether this change requires further changes of the dynamic sets $H_s(\mathbf{N})$, $H_d(\mathbf{N})$ and $H_u(\mathbf{N})$, and whether it requires an update of the admissible extensions of other intended applications. Any change may result in a series of further changes, which, in turn, may induce further changes and so on.

Obviously, for every revision of the present epistemic state by an application α, we can specify a tree that represents the operations that are in fact performed. This gives rise to Fig. 7.1. The present epistemic state is given by the sets $H_s(\mathbf{N})$, $H_d(\mathbf{N})$, $\mathbf{A}(\beta)$ (for any application β of a theory-element $\mathbf{T}$ in $\mathbf{N}$) as well as sets that represent the determinations among applications. Of course, one must wonder whether or not the TMS terminates on all possible inputs.

7.3.2 *Termination*

Termination has direct bearing on the complexity, soundness and completeness of the TMS. If the TMS does not terminate on some input, it does not make sense to determine the complexity class. For, a failure to terminate means that there are worst cases in which the TMS takes an infinite amount of running time. Moreover, failures of termination would imply that there are revision problems for which the algorithm gives no result at all. Correctness and completeness would then be violated trivially.

The following proof of termination makes use of the foundationalist principle that the epistemic rank of an intended application is given by the epistemic rank of the weakest intended application that determines it. First, we will show that certain kinds of infinite loop cannot occur. Second, that there is an upper finite bound on the length of a single branch in the operation tree. Third, that there is an upper finite bound on the number of child nodes of a single operation node in the tree. From this it follows that the computation of revisions by the present TMS will terminate on all possible inputs on condition that $|H(\mathbf{N})|$ is finite and $|EXT(\mathbf{T})(\mathbf{I}(\alpha)) \cap \mathbf{M}(\mathbf{T})|$ is finite for all intended applications α. Also, it must be assumed that the set $EXT(\mathbf{T})(\mathbf{I}(\alpha)) \cap \mathbf{M}(\mathbf{T})$ can be computed by a finite number of steps.

On the face of it, infinite loops are difficult to rule out because certain operations undo all effects of certain other operations. All operations effected by **revise**(α) will be undone in case α becomes rejected at a later stage. Hence, we must wonder whether the operation tree may contain, for example, a branch of the following kind:

$$\langle rev(\alpha), det(\beta_1), rev(\beta_1), det(\beta_2), rev(\beta_2), rej(\alpha)\rangle.$$

If such an order of operations were possible, we would run into an infinite loop because the rejection of α makes it necessary to undo all determinations induced by **revise**(α). This, however, results into unblocking α, i.e., into another call of **revise**(α), which may in turn lead to redoing the determinations of β_1 and β_2 so that α becomes rejected another time and so on. This particular sequence, however, is not a possibility for the following reasons. As β_2 is determined indirectly by α, its epistemic rank cannot be higher than that of α. More precisely, the members of $DI_w(\beta_2)$ cannot have a higher epistemic rank than the members of $DI_w(\alpha)$ at the stage when β_2 is determined by β_1. Hence, β_2 cannot reject α, even if the two are inconsistent with one another. In a similar vein, other kinds of infinite loops can be ruled out. Showing that certain kinds of sequences of operations cannot be repeated ad infinitum in a single branch, however, does not imply termination. There may still be infinite branches that do not contain an infinite repetition of a certain subsequence. However, we will also show that there is an upper finite bound on the length of a branch in the operation tree. Let us now work out the details.

Any branch in the operation tree that starts from the root has the following structure:

$$\langle rev(\alpha_1), op_{1,1}, \ldots, op_{1,k_1}, rev(\alpha_2), op_{2,1}, \ldots, op_{2,k_2}, rev(\alpha_3), \ldots\rangle \tag{7.1}$$

where $op_{i,j}$ stands for an operation other than **revise**.

Proposition 7.3 *Any branch of an operation tree of the TMS has the following property. If the operation* **revise**(α_{i+1}) *is the next revision operation that follows the operation* **revise**(α_i) *– in the sense of an operation sequence of type* (7.1)–, *then the epistemic rank of* α_{i+1} *is at most as high as that of* α_i.

By the epistemic rank of an intended application α we always mean the rank in its present state of determination. Hence, *epistemic rank of* α is shorthand for the epistemic rank of $\mathbf{T}(\psi)$, where ψ is some member of $DI_w(\alpha)$. Proposition 7.3 helps us establish another proposition concerning the structure of the operation tree:

Proposition 7.4 *There is no branch that has one of the following types as sub-branch:*

$$\langle \ldots, doubt(\alpha), \ldots, rev(\alpha), \ldots\rangle$$

$$\langle \ldots, rej(\alpha), \ldots, rev(\alpha), \ldots\rangle.$$

In other words, if an intended application α becomes rejected or doubted, **revise** will not be called another time for α. With the help of this proposition, we can show:

Proposition 7.5 *Any branch of an operation tree is finite.*

Proposition 7.6 *For any node of an operation tree, the number of immediate subnodes is smaller than or equal to* $2|H(\mathbf{N})|$.

Therefore:

Proposition 7.7 *Assume* $|H(\mathbf{N})|$ *is finite, and* $|EXT(\mathbf{T})(\mathbf{I}(\alpha)) \cap \mathbf{M}(\mathbf{T})|$ *is so for all intended applications* $\alpha \in H(\mathbf{N})$ *and the revision formula. Further, assume that the set* $EXT(\mathbf{T})(\mathbf{I}(\alpha)) \cap \mathbf{M}(\mathbf{T})$ *can be computed within a finite number of steps for all* $\alpha \in H(\mathbf{N})$ *and the revision formula. Then, the TMS (which is given by Algorithm* 7.4*) terminates for all revision formulas.*

This proposition follows from Propositions 7.5 and 7.6.

7.3.3 Computational Complexity

The input of a revision problem consists, in our TMS, of a set of formulas and sets of interpretations. As for the formulas, we have the set $H(\mathbf{N}) \cup E(\mathbf{N})$, where $E(\mathbf{N})$ designates the structuralist background theory. $H(\mathbf{N})$ is partitioned into the sets $H_s(\mathbf{N})$, $H_d(\mathbf{N})$ and $H_u(\mathbf{N})$. As for the interpretations, the TMS keeps track of the set $\mathbf{A}(\beta)$ of admissible extensions for any application β of any theory-element $\mathbf{T}$ in $\mathbf{N}$. The revision formula is given by a proposition of the form $b \in \mathbf{I}(\mathbf{T})$. The solution of the revision problem consists of a revised classification of the members of $H(\mathbf{N})$ into successful, doubted and unsuccessful intended applications. Furthermore, the sets $\mathbf{A}(\beta)$ are revised for all intended applications. We shall investigate how the number of computation steps depends on the syntactic size of $H(\mathbf{N})$. There is no need to explicitly consider the syntactic size of the sets $\mathbf{A}(\beta)$ ($\beta \in H(\mathbf{N})$). For, if the cardinality of the sets $\mathbf{A}(\beta)$ is bounded by a constant (as we shall assume), then the syntactic size needed to represent all sets $\mathbf{A}(\beta)$ ($\beta \in H(\mathbf{N})$) linearly depends on the cardinality of $H(\mathbf{N})$. There is no need to explicitly consider the syntactic size of $E(\mathbf{N})$ since $E(\mathbf{N})$ figures as an axiomatic background theory. If the maximal number of computation steps needed to solve a revision problem is bounded by a polynomial in n such that n represents the syntactic size of $H(\mathbf{N})$ (as we shall show), then this number of computation steps is bounded by a polynomial in n such that n represents the syntactic size of a string representing $H(\mathbf{N})$ and $E(\mathbf{N})$.

Let us now begin with determining the computational complexity of the initial procedure **revise_in**. Any revision with a proposition α starts by determining the admissible model extensions of α in **revise_in**. This procedure may (but does not necessarily) require consideration of all **T**-extensions of α, which are given by the set $EXT(\mathbf{T})(\mathbf{I}(\alpha))$. If the cardinality of both the descriptive vocabulary of **T** and the domain of interpretation of α is bounded by a constant, we can avoid running into the problem of combinatorial explosion. Under these conditions, we have controlled, local combinatorial explosion rather than global. For, it is easy to verify that computing

$$\sum_{\alpha \in H(\mathbf{N})} |EXT(\mathbf{T})(\mathbf{I}(\alpha))|$$

does not grow exponentially but only linearly with the cardinality of $H(\mathbf{N})$. If we were to consider, by contrast, global interpretations, i.e., interpretations not confined to theory-elements and intended applications thereof, their number would certainly grow exponentially with the syntactic size of the knowledge base. Hence, combinatorial explosion. $EXT(\mathbf{T})(\mathbf{I}(\alpha))$ depends exponentially on the cardinality of both the descriptive vocabulary of **T** and the domain of interpretation of α. These two cardinalities must therefore be assumed to be bounded by a constant.

Note, however, that the domain of interpretation is infinite when mathematical domains of interpretation are needed, such as natural, rational and real numbers. It goes without saying that both computers and human minds are usually unable to consider all members of $EXT(\mathbf{T})(\mathbf{I}(\alpha))$ when mathematical domains of interpretation are involved. The situation, however, is not hopeless. In the quantitative sciences, intended applications often have the wonderful property of having a unique model extension. (Recall that any measurement can be represented by means of an intended application that has this property.) In this case, it is often not necessary to consider all members of $EXT(\mathbf{T})(\mathbf{I}(\alpha))$ in order to determine the model extensions of α, i.e., the set $EXT(\mathbf{T})(\mathbf{I}(\alpha)) \cap \mathbf{M}(\mathbf{T})$. This is important since in the quantitative empirical sciences, new epistemic input consists, to a large extent, of new measurements.

In sum, tractability of the TMS can only be achieved if the cardinality of the set $EXT(\mathbf{T})(\mathbf{I}(\alpha))$ is bounded by a constant or the number of computation steps needed to determine the set $EXT(\mathbf{T})(\mathbf{I}(\alpha)) \cap \mathbf{M}(\mathbf{T})$ is bounded by a constant such that at least one of the two bounds holds for all $\alpha \in H(\mathbf{N})$. As we shall see more clearly in Sect. 7.3.4, these constants must be "reasonably small" so as to achieve efficient computation in the narrow sense.

Let us now move on to **revise**(α) and the other procedures of Algorithm 7.4. The number of computation steps that the TMS takes to execute **revise** crucially depends on two further quantities: $|EXT(\mathbf{T}(\alpha))(\alpha) \cap \mathbf{M}(\mathbf{T}(\alpha))|$ for each application $\alpha \in H(\mathbf{N})$ and $|H(\mathbf{N})|$. Let $CM(\alpha)$ designate the cardinality of the set of model extensions for an intended application α and n designate the cardinality of $H(\mathbf{N})$. Of importance is, furthermore, the number of links, which may be designated by m.

To see the impact of these quantities on the computational costs, recall that the TMS centres on the test of link satisfaction. Depending on the result of this test, operations of determination, doubt and rejection are induced, as explained above. Now, the test of link satisfaction for a pair of intended applications (α, β) involves computation steps that grow linearly with $CM(\alpha) \cdot CM(\beta) \cdot m$ in the worst case. Execution of **revise** requires, in the worst case, tests of link satisfaction for $n - 1$ intended applications. All commands other than the test of link satisfaction take either constant time or time that is linearly dependent on the cardinality of $H(\mathbf{N})$. Hence, we have:

$$Comp(\mathbf{revise}(\alpha)) = O(c_1 \cdot n + c_2 \cdot CM(\alpha) \cdot m \cdot \sum_{\beta \in H_s(\mathbf{N}) \cup H_d(\mathbf{N})} CM(\beta)) \tag{7.2}$$

where $Comp(\mathbf{revise}(\alpha))$ designates the number of computation steps. c_1 is a constant that accounts for the computation steps of commands other than the test of link satisfaction, such as the test of whether or not an intended application is a member of $H_s(\mathbf{N})$. c_2 accounts for the maximal costs of a single test of link satisfaction, i.e., the test whether or not two structures x_1 and x_2 satisfy a certain link $\mathbf{L}$. The *Big-O notation* used here is to be read as follows:

Definition 7.2 (Big-O notation) Let f and g functions $f, g : \mathbb{N} \rightarrow \mathbb{N}$. $f(n) = O(g(n))$ if and only if there are a constant c and a natural number n_0 such that for all natural numbers $n \geq n_0$, $f(n) \leq c \cdot g(n)$.

Assuming that $\overline{C}$ designates the arithmetic mean of the number of model extensions over all members of $H_s(\mathbf{N}) \cup H_d(\mathbf{N})$, (7.2) transforms to:

$$Comp(\mathbf{revise}(\alpha)) = O((c_1 + c_2 \cdot CM(\alpha) \cdot \overline{C} \cdot m) \cdot n) \tag{7.3}$$

where n designates the cardinality of $H(\mathbf{N})$. In the above discussion of **revise_in** it has been argued that the cardinality of the set of model extensions must be bounded by a constant for all intended applications. Let CM_m designate the maximal cardinality of this set. Then, we have:

$$Comp(\mathbf{revise}(\alpha)) = O((c_1 + c_2 \cdot CM_m \cdot \overline{C} \cdot m) \cdot n). \tag{7.4}$$

The operations **determine**, **doubt** and **reject** are inexpensive as compared to **revise**. Determination happens as part of the test of link satisfaction. **doubt** and **reject** are preceded by a comparison of the weakest intended applications that determine, respectively, the members of an inconsistent pair (α, β) of intended applications. As the priority ordering is modular, it suffices to compare one of the weakest applications that determine α with one of the weakest applications that determine β. This can be done in constant time. As for the division of applications into successful, doubted, and unsuccessful applications, we have the following worst-case scenarios: a substantial portion of $H(\mathbf{N})$ has to be moved from $H_s(\mathbf{N})$ to $H_d(\mathbf{N})$ in the case of **doubt**(α, β), and from $H_s(\mathbf{N})$ or $H_d(\mathbf{N})$ to $H_u(\mathbf{N})$ in the case of **reject**(α, β). Hence, we have:

$$Comp(\mathbf{doubt}(\alpha, \beta)) = O(c_3 \cdot n) \tag{7.5}$$

$$Comp(\mathbf{reject}(\alpha, \beta)) = O(c_4 \cdot n) \tag{7.6}$$

where c_3 and c_4 denote constants.

undetermine is more expensive than **determine** and **reject** since it may involve the (partial) redetermination of an intended application α after all determinations of α have been retracted. However, it is certainly not more expensive than **revise**. For this to be seen, note that the maximal number of calls of **satisfaction** within **revise** is greater than this number is for **undetermine**. And the computational complexity of both **revise** and **undetermine** is governed by the maximal number of calls of **satisfaction**. Finally, **reconsider** is a subprocedure of **undetermine**, which takes only constant time. Hence, we have:

$$Comp(\mathbf{undetermine}(\alpha)) \leq Comp(\mathbf{revise}(\alpha)). \tag{7.7}$$

We have thus established that each single operation in the tree takes at most a number of computation step that is linear in n and m. Shall we consider m, the number of links, as dependent on n? The number of links usually grows with the number of intended applications, for obvious reasons. The larger our knowledge base is, the larger is the number of concepts for which link satisfaction needs to be tested. Assuming a logarithmic or linear dependency of m on n is therefore reasonable. However, there is an obvious means to optimise the test of link satisfaction. For this test to be performed for a pair of intended applications (α, β), it suffices to test just those concepts that concern both the theory-element of α and that of β. For example, when testing link satisfaction for an intended application of **LP** (lever principle), the equality link for pH-value is of no concern. Since we have assumed that the number of concepts of a theory-element is bounded by a small constant, the number of links for which satisfaction is to be tested is independent of $H(\mathbf{N})$. Only for the sake of simplicity, we have written 'for all links **L** : ...' in the algorithm. An optimisation of this loop is not difficult to realise. Hence, we can assume m to be constant.

The crucial question that comes up at this point is, what is the maximum size of the operation tree in terms of single operations? If this tree involves, in the worst case, 2^n operations or more, the TMS algorithm would not be tractable. If, by contrast, the number of operations is bounded by a polynomial, tractability would be achieved.

To determine the complexity class of our TMS, we need to analyse the size of the operation tree. Let us first set forth a few propositions concerning the order and the number of revisions, doubts, rejections and determinations of a single tree.

Proposition 7.8 *If an operation* $\mathbf{revise}(\alpha_{i+1})$ *is executed after an operation* $\mathbf{revise}(\alpha)$*, the epistemic rank of* α_{i+1} *cannot be higher than that of* α*.*

Using this proposition, we can establish further propositions concerning the maximal size of an operation tree:

Proposition 7.9 *For all intended applications* α*, we can have at most one operation* $\mathbf{doubt}(\alpha)$ *or* $\mathbf{reject}(\alpha)$*, but not both operations, in one and the same operation tree.*

Proposition 7.10 *For all intended applications* α*, we can have at most* $(|H(\mathbf{N})|-1)$ *operations* ***undetermine***(α)*, in an operation tree.*

Proposition 7.11 *For all intended applications* α*, we can have at most* $(CM_m - 1)(|H(\mathbf{N})| - 1)$ *determinations of* α*, in an operation tree.*

Recall that CM_m is the maximal cardinality of the set of model extensions for any application in $H(\mathbf{N})$.

As for the number of revisions:

Proposition 7.12 *For all intended applications* α*, the number of operations* $\mathbf{revise}(\alpha)$*, in an operation tree, is smaller than or equal to* $(1 + CM_m)|H(\mathbf{N})|$.

These propositions help us analyse the computational costs for the complete execution of the operation tree:

(1) Revisions: $O(nComp(\mathbf{revise}))$
(2) Undeterminations: $O(nComp(\mathbf{undetermine}))$
(3) Rejections: $O(nComp(\mathbf{reject}))$
(4) Doubts: $O(nComp(\mathbf{doubt}))$.

(1) says that the computational (worst-case) complexity of the execution of all operations **revise** in the whole tree is $O(nComp(\textbf{revise}))$. (2), (3) and (4) are to be interpreted analogously. Note that determinations are brought about in a subroutine of **revise** so that no explicit consideration of their computational costs is needed.

From this analysis and the above estimation of the computational costs of **revise**, **doubt**, **reject** and **undetermine**, it follows that the TMS has in fact polynomial time complexity. To be more precise, Propositions 7.9–7.12 and Eqs. (7.4), (7.5), (7.6), and (7.7) let us infer the following theorem:

Theorem 7.1 *Let the TMS be given by Algorithm* 7.4. *Suppose there are constants* c_1 *and* c_2 *such that, for all applications* α *of* **N**, *(i)* $|EXT(\mathbf{T})(\alpha)|$ *is bounded by a constant* c_1, *or (ii)* $|EXT(\mathbf{T})(\alpha) \cap \mathbf{M}(\mathbf{T}(\alpha))|$ *is bounded by a constant* c_1 *and the computational costs of determining* $|EXT(\mathbf{T})(\alpha) \cap \mathbf{M}(\mathbf{T}(\alpha))|$ *is bounded by a constant* c_2. *Then, the computational complexity of determining belief changes by the TMS is* $O(n^2)$.

Hence, the TMS determines belief changes in polynomial time.

7.3.4 *Tractability and Modularity*

Let us review the steps leading to the present tractability result. We set out to analyse tractable reasoning using a modular approach. The notion of an application of a theory-element figures as the smallest modular unit in this approach. A theory-element **T** represents a general proposition, also called the *substantial law* of **T**. A theory-net is simply a set of theory-elements. In line with classical structuralism, we represent an application of a theory-element by a model-theoretic structure. More precisely, an application α stands for sentences of the logical form $b \in \mathbf{I}(\mathbf{T})$, where b is a model-theoretic structure. Likewise, the semantic consequences of applying a generalisation to a system of objects is represented by a set of model-theoretic structures, viz., the set $EXT(\mathbf{T})(\alpha) \cap \mathbf{M}(\mathbf{T}(\alpha))$.

In light of these facts, it is not surprising that the computational costs of determining belief changes crucially depend on four parameters. First, the number of applications within the theory-net **N**, i.e., the number of applications $b \in \mathbf{I}(\mathbf{T})$ such that $\mathbf{T} \in \mathbf{N}$. This parameter is designated by n in the above analysis of computational complexity. Second, the cardinality of the set $EXT(\mathbf{T})(\alpha)$, which represents the set of set-theoretic structures that may satisfy the substantial law of **T**. Third, the cardinality of the set $EXT(\mathbf{T})(\alpha) \cap \mathbf{M}(\mathbf{T}(\alpha))$, which represents the set of set-theoretic structures that actually satisfy the substantial law of **T**. Fourth, the number of computation steps needed, in the worst case, to determine $EXT(\mathbf{T})(\alpha) \cap \mathbf{M}(\mathbf{T}(\alpha))$.

Furthermore, it should not surprise us that our modular approach to tractability only succeeds if the size of all modular units is bounded by a constant. This is why we need to assume in Theorem 7.1 that $|EXT(\mathbf{T})(\alpha)|$ is bounded by a constant, or $|EXT(\mathbf{T})(\alpha) \cap \mathbf{M}(\mathbf{T}(\alpha))|$ and the computational costs of determining $|EXT(\mathbf{T})(\alpha) \cap \mathbf{M}(\mathbf{T}(\alpha))|$ are bounded by respective constants. Violation of this condition would mean that the computational costs of determining the semantic consequences of applying some generalisation to some system of entities is not bounded by a constant.

Admittedly, the size of the smallest modular units of reasoning must not only be bounded by a constant, but also "sufficiently small". As is well know, the notion of being sufficiently small is notoriously difficult to be made precise. However, at least for common sense generalisations, this assumption seems to be well justified. An application of such a generalisation, normally, concerns at most a handful of objects the properties of which are described without mathematical functions. Given these conditions, the sets $EXT(\mathbf{T})(\alpha)$ and $EXT(\mathbf{T})(\alpha) \cap \mathbf{M}(\mathbf{T}(\alpha))$ are bounded by "small constants". A case in point are simple laws about dispositional properties: bread nourishes a human being, metal conducts electricity, wood floats in water, etc.

Things are different for laws of scientific theories involving mathematical functions, paradigmatically functions mapping empirical objects to natural numbers, real numbers, or even vectors of such numbers. A case in point is Newtonian mechanics and the three-body problem. The latter problem is to determine the trajectory of three bodies that interact with one another through gravitational forces. The interactions between the Earth, the Sun and the Moon are the first historical instance of this problem. Set aside certain special cases of initial conditions, we need to use approximate, numerical approaches in order to (approximately) solve instances of this problem. Not surprisingly, the structuralist framework cannot help establish an exact solution. The set $EXT(\mathbf{T})(\alpha) \cap \mathbf{M}(\mathbf{T}(\alpha))$ cannot be represented using finite resources of memory for all applications α if space-time points are represented using real numbers. Even worse, the number of applications n is not finite. For, we need to consider a set of applications of Newton's equations and the law of gravitation such that this set captures a real-numbered interval of time points. Note, however, that approximate numerical solutions of an instance of the three-body problem are confined to considering a finite number of applications of Newton's equations and the law of gravitation. Moreover, such approximate solutions work with subsets of rational numbers instead of real numbers, thereby ensuring that the computational costs of determining $EXT(\mathbf{T})(\alpha) \cap \mathbf{M}(\mathbf{T}(\alpha))$, for a single application α, are bounded by a constant.

If the computational costs of reasoning within the smallest modular unit of reasoning are bounded by a constant, this implies that these costs do not grow with n, the number of applications of the theory-net $\mathbf{N}$. That is, the cardinality of the set $EXT(\mathbf{T})(\alpha)$ or the computational costs of determining the set $EXT(\mathbf{T})(\alpha) \cap \mathbf{M}(\mathbf{T}(\alpha))$ do not grow with n. For most laws of scientific theories as well as generalisations of everyday reasoning, the condition in question is satisfied and easy to justify. As suggested above, a single application of a generalisation used in everyday reasoning hardly ever concerns more than a handful of objects and properties. (We will discuss some exceptions below.) If so, the set $EXT(\mathbf{T})(\alpha)$ is bounded by a small constant, and does certainly not depend on n. More generally, the structuralist framework allows us to apply substantial laws to small subsystems of objects, while interdependencies between objects of different subsystems are captured by links or further applications. Hence, as long as the subsystems are small, it does not matter how many applications and how many substantial laws are in play within the theory-net.

Even for most laws of scientific theories, it is easy to see that the computational costs of determining the semantic consequences of a single application – i.e., the costs of determining $EXT(\mathbf{T})(\alpha) \cap \mathbf{M}(\mathbf{T}(\alpha))$ – does not depend on the number of applications in the theory-net **N**. Take Newton's law of gravitation. A single application of this law concerns exactly two spatiotemporal objects that attract one another. If we determine the gravitational forces using the law of gravitation with a degree of precision confined to subsets of rational numbers (as standardly done in numerical approaches), this determination does not depend on how many further objects exert further gravitational forces on our two objects. However, if we want to determine the trajectories of our two objects, we first need to calculate all gravitational forces and then calculate total forces that equal the sum of all component forces. So, the addition of component forces is a case where the number of applications may, at least, indirectly affect the size of the modular unit of reasoning.

Note, however, that in actual numerical calculations of classical trajectories the number of objects considered is very small compared to the number of applications of substantial laws. Trajectories calculated for our solar system take into account at most all planets and their moons as well as some smaller object, such as a satellite. Interactions between satellites as well as gravitational forces exerted by satellites on planets are ignored. But it is important to calculate the trajectories in a maximally possible fine-grained way, i.e., to take as many as computationally feasible space-time points into account. Each space-time point corresponds to a set of applications of substantial laws. Hence, for such numerical and approximate calculations of trajectories, the size of all modular units of reasoning remains bounded by a constant.

Arguably, it is a common strategy in scientific modelling to reduce the number of interacting objects so as to achieve tractability. In systems where such reductions are very difficult to achieve, our means of prediction and calculation are limited. As an example, take long-term weather predictions for a specific geographical region. A key challenge arising for predictions of economic systems is to find sensible reductions of the interacting agents and the relevant quantities considered.

Do we face similar challenges in everyday contexts? Admittedly, we do. Take certain general guidelines for major decisions, such as *maximise your happiness*, *help others as much as you can*, or *find a partner that ideally complements you*. Even if we could properly quantify happiness and utility, optimising these quantities involves an overwhelmingly large number of interrelated objects and quantities. From the perspective of the present approach to tractability, we conjecture that human agents cope with this complexity by somehow limiting the options and consequences considered so as to obtain computationally feasible, modular units of reasoning. Further research, including empirical investigations, needs to be done to confirm and to elaborate this conjecture.

Given there are constants c_1 and c_2 as described in Theorem 7.1, why is the computational complexity of the truth maintenance algorithm $O(n^2)$? In brief, we have shown that the number of computation steps needed, in the worst case, for a single suboperation of the TMS depends linearly on n. In this calculation, we have considered the following suboperations: revise, undetermine, rejection and doubt. Further, we have shown that the number of suboperations needed, in the worst case, to carry out a revision depends linearly on n too. Thus, we obtain $O(n^2)$ as computational complexity of the truth maintenance algorithm.

One further note on the role of the parameters c_1 and c_2 in Theorem 7.1 is in order. Strictly speaking, the theorem claims that the TMS algorithm is *fixed-parameter tractable*.[4] That is, the truth maintenance algorithm is tractable on condition of certain parameters being bounded by a constant. As explained, the parameters c_1 and c_2 are intended to capture upper bounds of the computational costs for a single modular unit of reasoning. We have argued that the assumption of such bounds is rather reasonable, at least for applications of substantial laws that are actually used by human agents and computers for calculations and inferences.

We understand the notion of tractability in the wider sense that includes fixed-parameter tractability, where qualifications concerning parameters remain important, of course. This is a reasonable convention since the notion of tractability aims to capture some notion of computational feasibility. If a problem is fixed-parameter tractable and the parameter in question is actually bounded by a small constant, then it is computationally feasible to solve non-trivial instances of the problem. As indicated above, some degree of vagueness is unavoidable when we argue whether or not a certain parameter is bounded by a "small constant". As a consequences of this, some degree of vagueness is involved in the question of whether it is computationally feasible to solve a fixed-parameter tractable problem.

Notably, at least some computer scientists understand the notion of tractability in the wider sense that includes fixed-parameter tractability. For example, Nebel [89, Sect. 4.1] shows that certain model-based belief revision algorithms are **P**-time complex for Horn logic if the revision formula is bounded by a constant. Likewise, the *logic of limited belief for reasoning with disjunctive information* by Liu et al. [80] is shown to be tractable. Strictly speaking, this logic is fixed-parameter tractable, and the parameter is given by how many levels of reasoning by cases are admitted in a proof.

7.4 Issues of Soundness and Completeness

7.4.1 The Inference Relation of the TMS

The intended use of the TMS consists in computing inferences from a structuralist default theory $\Theta(\mathbf{N})$. Let us make this inference relation precise:

---

[4] For a comprehensive introduction to fixed-parameter tractability and corresponding algorithms, see Flum and Grohe [54] and Niedermeier [91].

Explanation 7.2 (Inference relation of the TMS) Let $\Theta(\mathbf{N})$ be a structuralist default theory. Let, further, the TMS determine the set $H_s(\mathbf{N})$ and, for all $\alpha \in H(\mathbf{N})$, the sets $\mathbf{A}(\alpha)$. Then, we say that a sentence ϕ is inferred by the TMS if and only if there is some $\alpha \in H_s(\mathbf{N})$ such that ϕ is verified by all $x \in \mathbf{A}(\alpha)$.

This explanation is motivated by the semantics of theoretical terms (explained in Sect. 5.5), and the intended meaning of $H_s(\mathbf{N})$ and $\mathbf{A}(\alpha)$. Recall that the determination of $H_s(\mathbf{N})$ aims to capture the set of sentences $\alpha = \ulcorner y \in \mathbf{I}(\mathbf{T})\urcorner$ such that α is a member of any preferred subtheory of $\Theta(\mathbf{N})$. The determination of a set $\mathbf{A}(\alpha)$ is aimed at the set $\{x \mid (x, \mathbf{I}(\alpha)) \in AE(\mathbf{T})\}$. This gives rise to the following understanding of soundness and completeness of TMS:

Explanation 7.3 (Soundness and completeness of the TMS) The TMS is sound and complete if and only if, for all finite structuralist default theories $\Theta(\mathbf{N})$ and all revisions $\Theta(\mathbf{N}) *_i \alpha$ of such theories, the following two conditions are satisfied: (i) $H_s(\mathbf{N}') = \bigcap_{S \in P} S$, where P is the set of preferred subtheories of $\Theta(\mathbf{N}) \oplus_i \alpha$ and $\mathbf{N}'$ the theory-net extended by α. (ii) $\mathbf{A}(\alpha) = \{x \mid (x, \mathbf{I}(\alpha)) \in AE(\mathbf{T})\}$ for all $\alpha \in H_s(\mathbf{N}')$.

Once the sets $\mathbf{A}(\alpha)$ are determined, the test of whether or not some sentence ϕ is inferable from $\Theta(\mathbf{N})$ is straightforward. A structuralist default theory is finite if and only if $H(\mathbf{N})$ is finite and all domains of all intended applications are finite.

7.4.2 *Admissible* T*-Extensions*

In order to circumvent the problem of combinatorial explosion, the TMS has not attempted to find out whether or not there is a choice set $\mathbf{C}$ of the set

$$\mathbf{A}(\mathbf{N}) = \{B \mid \text{there is an } \alpha \in H_s(\mathbf{N}) \text{ s. t. } B = EXT(\mathbf{T})(\mathbf{I}(\alpha)) \cap \mathbf{M}(\mathbf{T})\}$$

such that for all $x_1, x_2 \in \mathbf{C}$, all links are satisfied. The set $\mathbf{A}(\mathbf{N})$ designates the set of the sets of model extensions for all intended applications α. The TMS rather tests whether, for all $\alpha \in H_s(\mathbf{N}) \cup H_d(\mathbf{N})$, all $x_1 \in \mathbf{A}(\alpha)$ and all $\beta \in H_s(\mathbf{N})$, there is an $x_2 \in \mathbf{A}(\beta)$ such that (x_1, x_2) satisfies all links. Hence, testing link satisfaction by the TMS is in fact based on the following postulate schema:

$$\begin{aligned}&\forall x_1 \forall y_1 \forall y_2 \exists x_2 ((x_1, y_1) \in \mathbf{AE}(\mathbf{T}_j) \wedge \\ &\quad (x_2, y_2) \in \mathbf{AE}(\mathbf{T}_k) \rightarrow \phi(\mathbf{L}_i)(x_1, x_2))\end{aligned} \qquad (\mathbf{P}_w 4(\mathbf{L}_i, \mathbf{T}_j, \mathbf{T}_k))$$

in place of the original

$$\begin{aligned}&\forall x_1 \forall y_1 \forall x_2 \forall y_2 ((x_1, y_1) \in \mathbf{AE}(\mathbf{T}_j) \wedge \\ &\quad (x_2, y_2) \in \mathbf{AE}(\mathbf{T}_k) \rightarrow \phi(\mathbf{L}_i)(x_1, x_2)).\end{aligned} \qquad (\mathbf{P}4(\mathbf{L}_i, \mathbf{T}_j, \mathbf{T}_k))$$

Let $\Theta_w(\mathbf{N})$ designate the structuralist default theory with $\mathbf{P}_w4$ in place of $\mathbf{P}4$. The question thus arises whether satisfaction of $\mathbf{P}_w4$ implies satisfaction of $\mathbf{P}4$. Unfortunately, the answer to this question is in the negative. For this to be seen, take the following example with propositional formulas[5]:

$$p \vee q \qquad (\alpha_1)$$

$$p \rightarrow s \qquad (\alpha_2)$$

$$q \rightarrow s. \qquad (\alpha_3)$$

In this example, the TMS fails to identify $\langle F, F \rangle$ as an inadmissible interpretation of α_2. That is, p and s cannot be both false in the application α_2. For, there are no admissible interpretations of α_1, α_2 and α_3 such that all links among α_1, α_2 and α_3 are satisfied, while p and s are assumed to be false. Rather, any admissible interpretation of α_2 and α_3 must determine s to be true, in the context of the admissible interpretations of the other applications. Hence, the TMS is an incomplete means of reasoning with respect to the structuralist belief revision scheme for default theories $\Theta(\mathbf{N})$.

It remains to investigate whether or not the TMS determines correctly the set $H_s(\mathbf{N})$ of successful intended applications. Also, we must wonder whether all inferences determined by the TMS are sound, while knowing that the TMS fails to be complete.

7.4.3 *Division of $H(\mathbf{N})$*

Unfortunately, the above example can be extended to construct cases where the TMS fails to determine correctly the set $H_s(\mathbf{N})$ of successful intended applications. Suppose α_1, α_2 and α_3 are extended by the following formulas:

$$u \vee v \qquad (\alpha_4)$$

$$u \rightarrow \neg s \qquad (\alpha_5)$$

$$v \rightarrow \neg s. \qquad (\alpha_6)$$

α_1–α_6 pass the test of link satisfaction if we do not have any further information about the interpretation of p, q, u, v and s. Hence, they would be classified as

[5]For simplicity, we work with propositional formulas and leave corresponding theory-elements implicit. Note that the propositional logical form of these formulas can be obtained by aptly chosen substantial laws. For example, the substantial law $\forall x(P(x) \vee Q(x))$ yields a formula of the form $p \vee q$ when applied to whatever intended applications. To be more precise: any application of this law is given by a structure $\langle D, P, Q \rangle$. If the sets P and Q of such a structure are empty, applying the law $\forall x(P(x) \vee Q(x))$ to this structure yields the claim that $P(a) \vee Q(a)$, where $D = \{a\}$.

successful, provided there are no other theory-elements. However, α_1–α_6 are clearly inconsistent. And this inconsistency is not detected by the TMS.

The example, furthermore, shows that reasoning with the TMS may not be sound. For, suppose the epistemic rank of α_1 is lower than that of α_2–α_6. Then, α_1 is not a member of any preferred subtheory of $\Theta(\mathbf{N})$, while the TMS would still infer that $p \vee q$. We have thus observed that reasoning with the TMS may not be sound and may not be complete.

The observed limitations of the TMS are rooted in a failure to account for reasoning by cases. Take the first example, given by the set of formulas $\{p \vee q, p \rightarrow s, q \rightarrow s\}$. Knowing that $p \vee q$, we know that at least one of the disjuncts must be true. Further, we know that in each case (p being true and q being true), we can infer s. But this type of disjunctive reasoning is not captured by the TMS. However, we can enable the TMS to carry out disjunctive reasoning by means of composite applications. Such applications will be described in Sect. 7.5.

7.4.4 Subsets of Doubted Applications

Another limitation of the TMS is that it tests for consistency between the revision formula and any doubted application, but not for consistency between the revision formula and sets of doubted applications. This works well for most cases – in the sense that an application α is accepted if and only if the revision formula $\alpha = \ulcorner b \in \mathbf{I}(\mathbf{T})\urcorner$ is a member of all preferred subtheories of $\Theta(\mathbf{N})$. There are scenarios, however, where this test of consistency is insufficient. Take the following example:

$$p \qquad (\beta_1)$$

$$p \rightarrow r \qquad (\beta_2)$$

$$r \rightarrow q \qquad (\beta_3)$$

$$\neg q. \qquad (\beta_4)$$

Suppose the TMS starts with an empty belief base. Then, this base is revised by β_1, β_2 and β_3 in this order. All these applications have the same epistemic rank. As a consequence of these revisions, β_1 determines β_2, which in turn determines β_3. Now, consider a further revision by β_4. As a consequence of this revision, β_3 and β_4 doubt one another, where β_3 is determined by β_1 and β_2. Hence, β_1 and β_2 are doubted as well. Since only successful applications can determine another application, β_1 cannot determine β_2 any more, after the revision by β_4. Nor can β_2 determine β_3. These determinations must therefore be retracted. Once they are retracted, the following revision formula will pass the consistency test with β_1, β_2, β_3 and β_4:

$$p \qquad (\alpha_1)$$

This means that α_1 would be accepted as successful, even if it has the same epistemic rank as $\beta_1, \beta_2, \beta_3$ and β_4. This is not as it should be since α_1 is not a member of all preferred subtheories of $\Theta = \langle\{\beta_1, \ldots, \beta_4, \alpha_1\}\rangle$.

To avoid this type of failure, the TMS would have to test for consistency of the revision formula with all consistent subsets of the set of doubted applications. Such a consistency test, however, would lead us directly to the problem of combinatorial explosion. Tractability could not be achieved any more if consistency between the revision formula and all proper subsets of the set of doubted applications was tested. However, by means of composite applications, we can enable the TMS to test consistency of the revision formula with sets of doubted applications. We shall describe composite applications in Sect. 7.5. Even with composite applications, the TMS remains tractable as long as the size of such applications is bounded by a constant.

7.4.5 *Irrelevant Determinations*

Suppose a successful application β is found inconsistent with the revision formula α, while α and β have the same epistemic rank. Then, the weakest applications determining β are doubted by α. This is a principle of the TMS. It is based on the assumption that all determinations of β are relevant for the inconsistency between α and β. For example, if γ_1 and γ_2 determine β, while α is inconsistent with β, then it is assumed that β would be consistent with α if it was not determined by γ_1 or not determined by γ_2. Hence, the applications γ_1 and γ_2 should be doubted, provided their epistemic rank is not higher than that of β. But the assumption that all applications determining an application β are relevant for an inconsistency is not always valid. Take the following example with propositional formulas:

$$p \qquad (\gamma_1)$$

$$w \qquad (\gamma_2)$$

$$(p \rightarrow r) \wedge (w \rightarrow s) \qquad (\beta)$$

$$\neg s. \qquad (\alpha)$$

Suppose γ_1, γ_2 and β are accepted, and all formulas in play have the same epistemic rank. Both γ_1 and γ_2 determine β. Then, the epistemic state is revised by α. Thus, β is found inconsistent with α, and so γ_1, γ_2, β and α are doubted. γ_1, however, is not relevant for the inconsistency between α and β. Rather, γ_1 is a member of all preferred subtheories of $\Theta = \langle T_1\rangle$, $T_1 = \{\gamma_1, \gamma_2, \beta, \alpha\}$. Hence, γ_1 should be classified as a successful application and not doubted by the TMS. We shall see in Sect. 8.4.1 that this problem can be solved by imposing a (moderate) constraint on the logical form of the substantial laws of the theory-elements in **N**.

7.4.6 *Weak Determinations*

Let us take a look at the following propositional formulas:

$$q \tag{α_1}$$

$$p \rightarrow q \tag{α_2}$$

$$\neg q. \tag{α_3}$$

Suppose α_1, α_2 and α_3 have the same epistemic rank. Further, suppose the empty epistemic state (in which only logical truths are believed) is revised by α_1, α_2 and α_3 in this order. Then, α_1 determines α_2. Now, assume that, when revising the epistemic state by α_3, α_3 is tested for consistency first with α_2 and then with α_1. In this scenario, α_1, α_2 and α_3 are moved from $H_s(\mathbf{N})$ to $H_d(\mathbf{N})$. But this behaviour is not correct because α_2 is a member of all preferred subtheories.

What went wrong? The determination of α_2 by α_1 is, in a certain sense, too weak to count as a proper determination. More precisely, this is a weak determination in the sense that there is no sentence that can be inferred from $\{\alpha_1, \alpha_2\}$ that could not be inferred from α_1 or α_2. We shall tackle the problem of weak determinations in Sect. 8.4.1.

7.4.7 *Preliminary Evaluation*

The price of tractable belief revision are potential failures of soundness and completeness. This is the lesson to be drawn from the preceding sections. The present observations concerning completeness and soundness may seem, at first sight, as discouraging as the complexity results of standard belief revision schemes. The situation, however, is not that hopeless. Our system should not be dismissed just on grounds of incompleteness and unsoundness.

First, it was too optimistic to envision the TMS to be sound and complete, even when assuming cardinality constraints on the descriptive vocabulary of theory-elements and the domains of intended applications. For, the problem of determining the intersection of the set of maximal consistent subsets of a set of formulas is **coNP**-complete if the mere test of consistency of a set of formulas is in **P** [89, Sect. 5]. Hence, if the TMS was sound and complete, we could infer from this and Theorem 7.1 that $\mathbf{P} = \mathbf{coNP}$. This would in turn imply $\mathbf{P} = \mathbf{NP}$. Moreover, if the TMS was sound and complete, it would give a polynomial algorithm for solving instances of 3-SAT. This problem consists in testing the satisfiability of propositional formulas in conjunctive normal form whose conjuncts have at most three literals. (We could simply split up such a formula into its conjuncts and take these conjuncts as theory-elements. In the case of 3-SAT, the size of the conjuncts are bounded by a constant, which is important for the tractability result of the TMS.)

Since 3-SAT is **NP**-complete [14, Chap. 2], soundness and completeness of the TMS would imply, then, that there is a polynomial algorithm for solving **NP**-complete problems. From this it would follow that $\mathbf{P} = \mathbf{NP}$. Unfortunately, the TMS fails to be sound and complete with respect to its intended belief revision scheme.

Second, both from a practical and cognitive point of view, a tractable algorithm that lacks completeness and soundness may be far more helpful than an intractable one that is complete and sound. It appears advisable, in this situation, to distinguish between a normative and a descriptive dimension according to which logical systems and belief revision schemes are to be judged. Using this distinction, we can say that $\Theta_w(\mathbf{N})$ (with its weakened link postulate) and the present TMS may score well in the descriptive dimension but are inferior to $\Theta(\mathbf{N})$ and standard belief revision schemes in the normative dimension.

Third, most importantly, reasoning with the TMS can be made to better approximate the norms of $\Theta(\mathbf{N})$ and the structuralist belief revision scheme by the formation of composite applications. In the next section, we shall explain an operation of composition on applications, which yields applications that are composed of other applications. Soundness and completeness will be achieved eventually for frame logic and a novel belief revision scheme in the final chapter of our investigation.

7.5 Approaching Soundness and Completeness

7.5.1 Composition

In object-oriented programming, there is an operation described as *composition* of classes. Given classes $C_1, \ldots, C_n$ have been defined with their methods and data fields, the programmer always has the option to introduce a composite class C_c of these classes (Sect. 2.2.2). In this composite class, methods and fields are definable that make use of the specific methods and fields of the component classes. The great benefit of composition is to have a means of reusing previously defined methods and fields in a structured and perspicuous way. Let us describe a corresponding operation for intended applications:

Explanation 7.4 (Composition of intended applications) Let $y_1, \ldots, y_n$ be intended applications of possibly different theory-elements $\mathbf{T}_1, \ldots, \mathbf{T}_n$. Some or even all of these applications may be compositions. Let $\alpha_1, \ldots, \alpha_n$ be the sentential representations of the applications $y_1, \ldots, y_n$. Then, we can construct a composite intended application y_c and a corresponding composite theory-element $\mathbf{T}_c$ (sententially represented by α_c) by means of the following steps:

(1) define a structure type $\langle D_1, \ldots, D_k, R_1, \ldots, R_m \rangle$ such that this structure type covers all base sets and relations that occur in the structure types of $\mathbf{T}_1, \ldots, \mathbf{T}_n$.

(2) Define $\mathbf{M}_p(\mathbf{T_c})$ as the conjunction of the definitions of $\mathbf{M}_p(\mathbf{T}_i)$ for all i, $1 \leq i \leq n$.
(3) Define $\mathbf{M}(\mathbf{T}_c)$ and $EXT(\mathbf{T}_c)$ as in step (2).
(4) Introduce postulates $\mathbf{P1}(\mathbf{T_c})$–$\mathbf{P6}(\mathbf{T_c})$, including link postulates between $\mathbf{T}_c$ and $\mathbf{T}_1, \ldots, \mathbf{T}_n$.
(5) Define α_c such that it encompasses the partial interpretations coming with $\alpha_1, \ldots, \alpha_n$.
(6) Define $DI_a(\alpha_c)$, $DI_p(\alpha_c)$ and $DI_a(\alpha_c)$ such that α_c inherits all determinations of its components.
(7) Extend the epistemic ordering $<$ such that the epistemic rank of α_c is equal to that of the weakest intended applications (or the weakest application) that α_c is composed of.
(8) If $\alpha_c \in H_s(\mathbf{N}) \cup H_d(\mathbf{N})$ is a composite application, then the test of consistency between α_c and the revision formula α is carried out after the test of consistency between α and all non-composite applications $\beta \in H_s(\mathbf{N}) \cup H_d(\mathbf{N})$ has been completed.

It is now easy to verify that the cases of incompleteness and unsoundness due to the TMS's failure to account for reasoning by cases can be resolved by means of composition. Take the example analysed in Sects. 7.4.2 and 7.4.3. Suppose we form a composite application of α_1, α_2 and α_3, thereby obtaining an application α_c with the following propositional representation

$$(p \vee q) \wedge (p \rightarrow s) \wedge (q \rightarrow s) \qquad (\alpha_c)$$

Then, the TMS will detect that $\langle F, T, F\rangle$ is not an admissible interpretation of $\langle p, q, s\rangle$. Since there are links to be satisfied between α_c and α_1–α_3, $\langle F, F\rangle$ will also be detected as an inadmissible interpretation of $\langle p, s\rangle$. And α_1 will be rejected in the context of α_2–α_6 if α_1 has a lower epistemic rank than α_2–α_6.

Likewise, we can test the consistency of the revision formula with subsets of doubted applications by means of composition. Thus, the problem observed in Sect. 7.4.4 disappears once we allow for compositions of applications. The problem of irrelevant determinations has already been addressed successfully by imposing a constraint on the logical form of substantial laws.

These observations gives rise to the conjecture that our TMS converges, by means of composition, to the eternal norms of classical logic, which govern the understanding of consistency in standard belief revision schemes. However, we need to make sure that compositions are non-vacuous. That is, we need to ensure that any member of a composite application is relevant for the determinations of the composition that are additional to the determinations of its components. This requirement will be explicitly introduced in the next chapter (Sect. 8.5.2).

Composition is thus a means to approach soundness and completeness of the TMS. The price we have to pay for allowing composition of intended applications is an increase in the computational costs. Composition leads to more involved intended applications whose domains of interpretation and whose descriptive vocabulary is likely to exceed the bounds that had to be assumed for the above tractability result. Yet, an important merit of compositions is that we can gradually accept higher computational challenges, depending on how much machine and human cognitive power is available, respectively. Composition does not mean that everything must be composed with everything else. Quite to the contrary, we need to look for cases that promise to give new, interesting and more accurate determinations. Also, an important purpose of compositions is to yield intended applications that either have the favourable property of having a unique model extension or can be determined to gain this property.

In the next chapter, we shall further pursue the objective of tractable belief revision, building upon the results of the present chapter. In particular, we shall revise our understanding of links in such a manner that link satisfaction can be tested completely in polynomial time. This revised understanding of links gives rise to a new logic for which a sound and complete proof theory can be given. The rationale of developing such a logic is to adjust our logical theory of reasoning to the cognitive and computational constraints of tractability.

7.5.2 More Defeasible Reasoning

Composition fits very nicely into the overall rationale of defeasible reasoning. Recall that this kind of reasoning is ampliative, risky and adventurous since its inferences may be defeated at a later stage of information. Both everyday and scientific reasoning exhibit such inferences because it would be impractical, if not even impossible, to base empirical science and everyday planning on conservative inferences. It seems impractical and impossible, in a similar vein, to perform belief revision and nonmonotonic reasoning by algorithms that guarantee completeness and soundness but lack tractability.

Arguably, there are two kinds of risk one takes when exercising reasoning in science and everyday life. First, inferences may be defeated at a later stage of inquiry through new information or new theories. Second, inferences may be defeated through more accurate computations. While standard approaches to nonmonotonic reasoning have focused on the first kind of risk, we have made an attempt at accounting for both. Potentially more accurate computations are induced, in our system, through the composition of intended applications. We can even view composition as a particular kind of epistemic input and write

$$\Theta_w(\mathbf{N}) * C(\alpha_1, \ldots, \alpha_n)$$

when the epistemic state is revised by the composition of the intended applications $\alpha_1, \ldots, \alpha_n$.

Chapter 8
Frame Logic

In the previous two chapters, we have described tractable reasoning with frames by an axiomatic system and an algorithm that performs operations on set-theoretic structures. But the TMS only approximately determines belief changes in the structuralist framework. And we are lacking yet a proper logic of frames that is tractable and furnished with a fully fledged proof theory. We shall now devise such a logic using the tractability result of the truth maintenance system. Thus, we tailor a logic to meet the cognitive and computational constraints of tractability.

8.1 Revising Our Understanding of Links

A major obstacle to a sound and complete determination of belief changes has been the problem that link satisfaction could not be tested completely in polynomial time by the TMS. In light of this observation, let us consider a revised understanding of links:

Explanation 8.1 (Link) There is a link between two applications α and β if and only if there is an atomic sentence $R(c_1, \ldots, c_n)$ such that (i) $R(c_1, \ldots, c_n)$ is a member of the sublanguages of both α and β, and (ii) the truth value of $R(c_1, \ldots, c_n)$ is determined by the admissible extensions of α or β, or both.

Links, thus, concern atomic sentences that are interpreted by two applications and that are semantically determined by the admissible extensions of at least one application. We say that an atomic sentence is determined by an application if and only if it is verified by all admissible extensions of at least one application or falsified by all admissible extensions of at least one application. If $R(c_1, \ldots, c_n)$ is determined to be true by α and forms a link between α and β, then this link constrains the admissible extensions of β such that $R(c_1, \ldots, c_n)$ must be verified by all admissible extensions of β. There is an inconsistency between α and β if and

H. Andreas, *Dynamic Tractable Reasoning*, Synthese Library 420,
https://doi.org/10.1007/978-3-030-36233-1_8

only if there is an atomic sentence $R(c_1, \ldots, c_n)$ that is determined to be true by the admissible extensions of α and determined to be false by the admissible extensions of β, or vice versa. These two cases amount to a failure of link satisfaction.

We can express this revised understanding of links by revising the link postulate schema $P4(\mathbf{T})$ as follows:

$$\begin{aligned}&\forall x \forall y \forall y'((x, y) \in AE(\mathbf{T}) \wedge (c_1, \ldots, c_n) \in (R)_x \wedge \\ &\quad \exists x'(x' \in EXT(\mathbf{T}')(y') \wedge (c_1, \ldots, c_n) \in (R)_{x'}) \rightarrow \\ &\quad \forall x'((x', y') \in AE(\mathbf{T}') \rightarrow (c_1, \ldots, c_n) \in (R)_{x'})).\end{aligned} \qquad (\mathbf{P4.1(T)})$$

This postulate schema simply says that, if all admissible **T**-extensions of an intended application y are such that they verify $R(c_1, \ldots, c_n)$ and if there is a $\mathbf{T}'$-extension of an intended application y' that verifies $R(c_1, \ldots, c_n)$, then all admissible $\mathbf{T}'$-extensions of y' must verify $R(c_1, \ldots, c_n)$. (n, simply, stands for the arity of the relation R.) Likewise, if an atomic sentence is determined to be false by the admissible extensions of an application α, but verified by some $\mathbf{T}'$ extension of an application β, then all admissible extensions of β must falsify the sentence in question:

$$\begin{aligned}&\forall x \forall y \forall y'((x, y) \in AE(\mathbf{T}) \wedge (c_1, \ldots, c_n) \notin (R)_x \wedge \\ &\exists x(x \in EXT(\mathbf{T})(y) \wedge (c_1, \ldots, c_n) \in (R)_x) \rightarrow \\ &\forall x'((x', y') \in AE(\mathbf{T}') \rightarrow (c_1, \ldots, c_n) \notin (R)_{x'})).\end{aligned} \qquad (\mathbf{P4.2(T)})$$

Note that applying and testing the satisfaction of these postulate schemas is computationally feasible insofar as it does not involve combinatorial explosion of computation steps as long as the cardinality of the set of model extensions of the application α is bounded by a constant. On the original understanding of links, by contrast, a complete test of link satisfaction requires consideration of all the choice sets **C** of the set

$$\mathbf{A(N)} = \{B \mid \text{there is an } \alpha \in H_s(\mathbf{N}) \text{ s. t. } B = EXT(\mathbf{T})(\mathbf{I}(\alpha)) \cap \mathbf{M(T)}\}.$$

This test leads to the infamous combinatorial explosion of computation steps, in the worst case. Hence, testing satisfaction of **P**4.1(**T**) and **P**4.2(**T**) is tractable, while testing satisfaction of $P4(\mathbf{T})$ is not. How can we exploit this result for devising a proper logic of frames? Such a logic would be founded upon a strict distinction between modular reasoning within the application of a frame and transmodular reasoning between applications of frames:

(1) Within the application of a frame concept, full first-order logic remains valid.
(2) Only atomic sentences and negations thereof can be inferred from one application of a frame concept to another.

At the core of frame logic is an inference rule that links applications of frame concepts with one another. Let

$$R(c_1, \ldots, c_n)|_\alpha \qquad \textit{and} \qquad \neg R(c_1, \ldots, c_n)|_\alpha$$

denote, respectively, that $R(c_1, \ldots, c_n)$ and $\neg R(c_1, \ldots, c_n)$ is asserted within the application α. Our inference rule can now be expressed as follows:

$$\frac{R(c_1, \ldots, c_n)|_\alpha}{R(c_1, \ldots, c_n)|_\beta}$$

$$\frac{\neg R(c_1, \ldots, c_n)|_\alpha}{\neg R(c_1, \ldots, c_n)|_\beta}$$

where α and β are schematic symbols that designate intended applications.

8.2 Frame Logic

8.2.1 Syntax

Recall that applications of theory-elements serve as modular units of reasoning in our analysis of thinking in frames. Further, we have assumed that, for any application α, there is a language $L(V(\alpha))$. $V(\alpha)$ designates the descriptive vocabulary, comprising relation symbols and individual constants. $L(V(\alpha))$ designates the set of well-formed formulas of α. Although language remains implicit in classical structuralism, we have demonstrated that it is possible and fruitful to reintroduce syntactic methods into the structuralist framework. We shall now spell out the syntax of frame logic.

Definition 8.1 (Alphabet of $L_F(\mathbf{N})$) Let $L_F(\mathbf{N})$ be a frame language associated with a theory-net **N**. The alphabet of this language consists of

(1) a set N_a of names for applications of theory-elements
(2) symbols for sets of applications
(3) all logical symbols of first-order logic, including identity but without function symbols
(4) for any $\alpha \in N_a$, the symbols of $V(\alpha)$.[1]

[1]Inclusion of function symbols is not difficult to achieve. In not representing them, we merely follow the structuralist convention to represent functions by many-to-one relations.

Well-formed formulas of $L_F(\mathbf{N})$ are built as follows:

Definition 8.2 (Well-formed formula of $L_F(\mathbf{N})$) Let $L_F(\mathbf{N})$ be a frame language associated with a theory-net **N** and a corresponding alphabet.

(1) If $\phi \in L(V(\alpha))$ and $\alpha \in N_a$, then $\phi|_\alpha$ is a formula of $L_F(\mathbf{N})$.
(2) Let A be a set of applications. If, for all $\alpha \in A$, $\phi \in L(V(\alpha))$ and $\alpha \in N_a$, then $\phi|_A$ is a formula of $L_F(\mathbf{N})$.
(3) Nothing else is a formula of $L_F(\mathbf{N})$.

The core idea of the frame language $L_F(\mathbf{N})$ is to relativise formulas to applications, which then serve as modular units of reasoning. This will allow us to explicate the interplay of intramodular and transmodular reasoning. Certain auxiliary formulas will be needed as well, as we shall see when we introduce natural deduction rules for frame logic. These auxiliary formulas are formulated in terms of set-theoretic symbols and the descriptive vocabulary of a theory-net **N**, such as symbols for theory-elements, substantial laws, etc.

Why do we include formulas that are relativised to a set A of applications as opposed to a single application? Now, besides making assertions about specific applications, we also want to make general assertions that we think hold for a range of applications. Also, we want to be able to draw conclusions from such assertions.

8.2.2 Natural Deduction Rules

Let us now move on to the natural deduction rules of frame logic. These rules extend and modularise the inference rules of a natural deduction system for standard first-order logic. In essence, we relativise formulas to applications so as to enable a corresponding relativisation of propositions: any proposition is asserted relative to an instance of a frame or a set of frame instances. Within one and the same frame instance, full first-order logic remains valid. What follows are the natural deduction rules for the propositional connectives[2]:

$$\frac{\phi_1|_\alpha \ldots \phi_n|_\alpha}{\phi_1 \wedge \ldots \wedge \phi_n|_\alpha} \qquad \text{(And Introduction)}$$

$$\frac{\phi_1 \wedge \ldots \wedge \phi_n|_\alpha}{\phi_i|_\alpha} \quad [1 \leq i \leq n] \qquad \text{(And Elimination)}$$

[2]The formulation of the following rules draws on the exposition of first-order logic in Zach [122] and Genesereth and Kao [63].

$$\frac{\phi_i|_\alpha}{\phi_1 \vee \ldots \vee \phi_n|_\alpha} \qquad [1 \leq i \leq n] \qquad \text{(Or Introduction)}$$

$$\frac{\begin{array}{c}\phi_1 \vee \ldots \vee \phi_n|_\alpha \\ \phi_1|_\alpha \vdash_s \psi|_\alpha \; \ldots \; \phi_n|_\alpha \vdash_s \psi|_\alpha\end{array}}{\psi|_\alpha} \qquad \text{(Or Elimination)}$$

where $\phi \vdash_s \psi$ means that there is a subproof of ψ from ϕ.

$$\frac{\phi|_\alpha \vdash_s \psi|_\alpha \qquad \phi|_\alpha \vdash_s \neg\psi|_\alpha}{\neg\phi|_\alpha} \qquad \text{(Negation Introduction)}$$

$$\frac{\neg\neg\phi|_\alpha}{\phi|_\alpha} \qquad \text{(Negation Elimination)}$$

$$\frac{\phi|_\alpha \vdash_s \psi|_\alpha}{\phi \to \psi|_\alpha} \qquad \text{(Implication Introduction)}$$

$$\frac{\phi|_\alpha \qquad \phi \to \psi|_\alpha}{\psi|_\alpha}. \qquad \text{(Implication Elimination)}$$

The rules for the quantifiers and identity are likewise intuitive:

$$\frac{\phi(a)|_\alpha}{\forall x\phi(x)|_\alpha} \qquad \text{(Universal Introduction)}$$

[where a does not occur in any active assumption of $\phi(a)$]

$$\frac{\forall x\phi(x)|_\alpha}{\phi(t)|_\alpha} \qquad [t \text{ is a ground term of } L(V(\alpha))] \qquad \text{(Universal Elimination)}$$

$$\frac{\phi(t)|_\alpha}{\exists x\phi(x)|_\alpha} \qquad \text{(Existential Introduction)}$$

$$\frac{\exists x\phi(x)|_\alpha \qquad \phi(a)|_\alpha \vdash_s \psi|_\alpha}{\psi|_\alpha} \qquad \text{(Existential Elimination)}$$

[where a does not occur

in any active assumption of ψ]

$$\frac{}{t = t|_\alpha} \qquad [t \text{ is a term of } L(V(\alpha))] \qquad \text{(Equality Introduction)}$$

$$\frac{\phi(t_1)|_\alpha \qquad t_1 = t_2|_\alpha}{\phi(t_2)|_\alpha} \qquad \text{(Equality Elimination)}$$

[where t_1 and t_2 are ground terms of $L(V(\alpha))$].

Such are the natural deduction rules for propositions that are relativised to a single intended application. Obviously, these rules differ from their originals in a standard system of natural deduction merely in the relativisation to an application.

We shall now introduce inference rules for propositions that are relativised to a set of applications. Strictly speaking, these rules are redundant since they can be derived from the just introduced rules together with the inference rules Scope Restriction, Instantiation, Comprehension and Extension to be introduced below. However, specific inference rules for propositions relativised to a set of applications help us formalise and understand reasoning with modular units. Our objective is not to maximise simplicity, but to give an intuitive and cognitively more refined representation of reasoning.

$$\frac{\phi_1|_A \ldots \phi_n|_A}{\phi_1 \wedge \ldots \wedge \phi_n|_A} \qquad \text{(And Introduction)}$$

$$\frac{\phi_1 \wedge \ldots \wedge \phi_n|_A}{\phi_i|_A} \qquad [1 \leq i \leq n] \qquad \text{(And Elimination)}$$

$$\frac{\phi_i|_A}{\phi_1 \vee \ldots \vee \phi_n|_A} \qquad [1 \leq i \leq n] \qquad \text{(Or Introduction)}$$

$$\frac{\begin{array}{c}\phi_1 \vee \ldots \vee \phi_n|_A \\ \phi_1|_A \vdash_s \psi|_A \ \ldots \ \phi_n|_A \vdash_s \psi|_A\end{array}}{\psi|_A} \qquad \text{(Or Elimination)}$$

$$\frac{\phi|_A \vdash_s \psi|_A \qquad \phi|_A \vdash_s \neg\psi|_A}{\neg\phi|_A}$$ (Negation Introduction)

$$\frac{\neg\neg\phi|_A}{\phi|_A}$$ (Negation Elimination)

$$\frac{\phi|_A \vdash_s \psi|_A}{\phi \rightarrow \psi|_A}$$ (Implication Introduction)

$$\frac{\phi|_A \qquad \phi \rightarrow \psi|_A}{\psi|_A}$$ (Implication Elimination)

$$\frac{\phi(a)|_A}{\forall x\phi(x)|_A}$$

[where a does not occur in any active assumption of $\phi(a)$] (Universal Introduction)

$$\frac{\forall x\phi(x)|_A}{\phi(t)|_A}$$

[where, for all $\alpha \in A$, t is a ground term of $L(V(\alpha))$] (Universal Elimination)

$$\frac{\phi(t)|_A}{\exists x\phi(x)|_A}$$ (Existential Introduction)

$$\frac{\exists x\phi(x)|_A \qquad \phi(a)|_A \vdash_s \psi|_A}{\psi|_A}$$

[where a does not occur in any active assumption of ψ] (Existential Elimination)

$$\frac{}{t = t|_A} \qquad \text{(Equality Introduction)}$$

[where t is a term of $L(V(\alpha))$ for all $\alpha \in A$]

$$\frac{\phi(t_1)|_A \qquad t_1 = t_2|_A}{\phi(t_2)|_A} \qquad \text{(Equality Elimination)}$$

[where t_1 and t_2 are ground terms of $L(V(\alpha))$ for all $\alpha \in A$].

Now, it is time to work out the inference rules that are distinctive of reasoning with frames. Let us begin with inference rules that govern the scope of an assertion in terms of applications and sets of applications:

$$\frac{\phi|_A}{\phi|_B} \quad [B \subseteq A] \qquad \text{(Scope Restriction)}$$

$$\frac{\phi|_A}{\phi|_\alpha} \quad [\alpha \in A] \qquad \text{(Scope Instantiation)}$$

$$\frac{\phi|_\alpha}{\phi|_{\{\alpha\}}} \qquad \text{(Scope Comprehension)}$$

$$\frac{\phi|_A \qquad \phi|_B}{\phi|_C} \quad [C = A \cup B]. \qquad \text{(Scope Extension)}$$

Recall that any admissible extension of an intended application of a theory-element $\mathbf{T}$ must satisfy the substantial law of $\mathbf{T}$. Hence, we introduce inference rules that allow us to infer the substantial law of the theory-element of an intended application:

$$\frac{}{\phi|_A} \quad [\mathbf{T}(\alpha) = \phi \text{ for all } \alpha \in A] \qquad \text{(Law Application 1)}$$

$$\frac{}{\phi|_\alpha} \quad [\mathbf{T}(\alpha) = \phi]. \qquad \text{(Law Application 2)}$$

We do not consider the condition $\mathbf{T}(\alpha) = \phi$ an explicit premise because this equation is not part of the object language $L_F(\mathbf{N})$. Hence, the condition is put in square brackets. Similar considerations apply to the conditions in square brackets of the above rules Scope Restriction, Instantiation, Comprehension and Extension.

All of the preceding rules formalise intramodular reasoning. These are the inference rules of transmodular reasoning:

$$\frac{R(c_1, \ldots, c_n)|_\alpha}{R(c_1, \ldots, c_n)|_\beta} \qquad \text{(Link Exploitation 1)}$$

[where $R(c_1, \ldots, c_n) \in L(V(\beta))$]

$$\frac{\neg R(c_1, \ldots, c_n)|_\alpha}{\neg R(c_1, \ldots, c_n)|_\beta} \qquad \text{(Link Exploitation 2)}$$

[where $R(c_1, \ldots, c_n) \in L(V(\beta))$].

We have thus come into a position to define the proof theory of frame logic:

Definition 8.3 ($\Gamma \vdash_F \phi$) Let Γ be a set of $L_F(\mathbf{N})$ sentences and let ϕ be such a sentence. $\Gamma \vdash_F \phi$ if and only if there is a tree of $L_F(\mathbf{N})$ sentences such that

(1) the topmost sentences are either in Γ or premises of a completed subproof. A subproof is complete if and only if it has been used as a justification of an application of an inference rule.
(2) The bottommost sentence is ϕ.
(3) Every sentence in the tree, except ϕ, is a premise of an application of an inference rule (of natural deduction frame logic) such that the conclusion of this application stands directly below that sentence.

In sum, the inference rules of natural deduction frame logic are given by the standard natural deduction rules relativised to applications and sets of applications, and the inference rules Scope Restriction, Scope Instantiation, Scope Comprehension, Scope Extension, Law Application 1 and 2, Link Exploitation 1 and 2.

8.2.3 Semantics

For a given theory-net $\mathbf{N}$, we now have two formal languages. First, $L(\mathbf{N})$, which contains the set-theoretic descriptive terms $\mathbf{AE(T)}$, $\mathbf{I(T)}$, $\mathbf{M(T)}$, etc., for any $\mathbf{T} \in \mathbf{N}$. Second, the frame language $L_F(\mathbf{N})$, as just introduced. The two languages have different, albeit related notions of satisfaction for formulas. The semantics of $L(\mathbf{N})$ is given by the semantics of theoretical terms, as explained in Sect. 5.5. We shall now define the semantics of $L_F(\mathbf{N})$ sentences, which rests on the semantics of $\mathbf{L(N)}$.

Definition 8.4 (Semantics of $L_F(\mathbf{N})$ sentences) ν_F: $L_F(\mathbf{N})^S \to \{T, F, I\}$, where $L_F(\mathbf{N})^S$ designates the set of $L_F(\mathbf{N})$ formulas that are sentences. Let $\mathbf{AE(T)}$ be interpreted by some $\mathbf{L(N)}$ structure $\mathcal{A}$.

(1) $\nu_F(\phi|_\alpha) = T$ iff for all x such that $(x, \mathbf{I}(\alpha)) \in \mathbf{AE}(\mathbf{T}(\alpha))$, $x \models \phi$.
(2) $\nu_F(\phi|_A) = T$ iff for all $\alpha \in A$ and for all x, if $(x, \mathbf{I}(\alpha)) \in \mathbf{AE}(\mathbf{T}(\alpha))$, then $x \models \phi$.
(3) $\nu_F(\phi|_\alpha) = F$ iff for all x such that $(x, \mathbf{I}(\alpha)) \in \mathbf{AE}(\mathbf{T}(\alpha))$, $x \models \neg\phi$.
(4) $\nu_F(\phi|_A) = F$ iff for all $\alpha \in A$ and for all x, if $(x, \mathbf{I}(\alpha)) \in \mathbf{AE}(\mathbf{T}(\alpha))$, then $x \models \neg\phi$.
(5) $\nu_F(\phi|_\alpha) = I$ iff neither $\nu(\phi|_\alpha) = T$ nor $\nu(\phi|_\alpha) = F$.
(6) $\nu_F(\phi|_A) = I$ iff neither $\nu(\phi|_A) = T$ nor $\nu(\phi|_A) = F$.

I stands for the indeterminate value. These truth rules rest on a given interpretation of the theoretical terms $\mathbf{AE}(\mathbf{T})$ for all $\mathbf{T} \in \mathbf{N}$. As the truth rules of $L_F(\mathbf{N})$ sentences rest on an interpretation of $\mathbf{AE}(\mathbf{T})$, we can say what it is for an $L(\mathbf{N})$ structure to satisfy an $L_F(\mathbf{N})$ sentence:

Definition 8.5 ($\mathcal{A} \models_F \phi$) Let ϕ be an $L_F(\mathbf{N})$ sentence and $\mathcal{A}$ an $L(\mathbf{N})$ structure. $\mathcal{A}$ satisfies ϕ – in symbols: $\mathcal{A} \models_F \phi$ – if and only if $\nu_F(\phi) = T$ on the interpretation of $\mathbf{AE}(\mathbf{T})$ by $\mathcal{A}$.

$\mathbf{AE}(\mathbf{T})$ and $\mathbf{I}(\mathbf{T})$ are, as it were, externally interpreted. That is, these terms are not part of the descriptive vocabulary of $L_F(\mathbf{N})$, even though they define its semantics. The reason for this is that the division into frames and frame instances is externally given. Our frame logic accounts for reasoning within such a division but not about such a division. Hence, the division into frames and frame instances is assumed to be fixed. It can be expanded and revised by means of AGM-style operations of belief revision, as shown in Chap. 6. Moreover, the division into frames can be changed by means of composition of intended applications.

Recall that we introduced the semantics of theoretical terms in such a manner that there is a range of admissible interpretations of a language $L(V)$ with theoretical terms. And this range is constrained by a set of postulates interpreting the theoretical terms. $\mathbf{AE}(\mathbf{T})$, thus, is interpreted by $P(\mathbf{N})$, the structuralist system of postulates, as expounded in Sect. 5.4. Note, however, that in order to achieve tractability, we revised our understanding of links by replacing $P4(\mathbf{T})$ with the weaker postulates $P4.1(\mathbf{T})$ and $P4.2(\mathbf{T})$. Hence, in what follows we assume that $\mathbf{AE}(\mathbf{T})$ is interpreted by $P_F(\mathbf{N})$, which results from $P(\mathbf{N})$ by replacing $P4(\mathbf{T})$ with the postulates $P4.1(\mathbf{T})$ and $P4.2(\mathbf{T})$. We now are in a position to define the relation of logical consequence for frame logic:

Definition 8.6 ($\Gamma \models_F \phi$) Let Γ be a set of $L_F(\mathbf{N})$ sentences and let ϕ be such a sentence. Let $\mathbf{S}_a$ be the set of admissible structures of $L(\mathbf{N})$, as defined by Definition 5.7, where $P(\mathbf{N})$ is replaced by $P_F(\mathbf{N})$. Also, we assume that any relation R of an intended application $\mathbf{I}(\alpha)$ is (partially) interpreted by the empty set. ϕ is a frame-logical consequence of Γ – in symbols: $\Gamma \models_F \phi$ – if and only if, for all $\mathcal{A} \in \mathbf{S}_a$, if $\mathcal{A} \models_F \alpha$ for all $\alpha \in \Gamma$, then $\mathcal{A} \models_F \phi$.

Why do we assume that any relation R of an intended application $\mathbf{I}(\alpha)$ is (partially) interpreted by the empty set? If we did not make this assumption, then any $L_F(\mathbf{N})$ sentence that is true in virtue of the partial interpretation by $\mathbf{I}(\alpha)$ would be a frame-logical consequence of the empty set. This would give us an awkward notion of a logic, accepting too many consequences as logically valid. Instead of representing singular facts by partial interpretations, we can easily represent such facts by sentences $\beta|_\alpha$, where β is a literal and $\alpha \in N_a$.

Note, furthermore, that $\models_F$ is always relative to a set of postulates $P_F(\mathbf{N})$ and a division of our domain of interpretation into intended applications. Hence, the empty set has some non-trivial frame-logical consequences, such as $\lambda|_\alpha$ if λ is the substantial law of $\mathbf{T}(\alpha)$. In this respect, $\models_F$ resembles Carnap's notion of a *P-consequence* in his *Logical Syntax of Language* [30].[3]

Note, finally, that the semantics of an $L_F(\mathbf{N})$ sentence is doubly supervaluationist: for an $L_F(\mathbf{N})$ sentence $\phi|_\alpha$ to be true, ϕ must be verified by all x such that $(x, \mathbf{I}(\alpha)) \in \mathbf{AE}(\mathbf{T}(\alpha))$, and this condition must hold on all admissible interpretations of $L(\mathbf{N})$.

8.2.4 Soundness and Completeness

Frame logic can be shown to be sound in a relatively straightforward manner.

Theorem 8.1 (Soundness) *Let Γ be a set of $L_F(\mathbf{N})$ sentences and let ϕ be such a sentence. If $\Gamma \vdash_F \phi$, then $\Gamma \models_F \phi$.*

The proof is given in the Appendix. When it comes to completeness, we encounter the following dilemma. On the one hand, certain constraints of finiteness need to be assumed in order to show completeness, at least for the quantifier-free fragment of frame logic. In particular, we shall assume that, for any application α, the domain D of α has finite cardinality. On the other hand, we know that the constraint to finite domains rules out the possibility of an axiomatisation. The problem is that, once we assume that any domain of any application is finite, we run into the problem that finite first-order logic cannot be axiomatised.[4] The problem of whether frame logic can be shown to be complete on weaker than the below specified constraints must be left for future research.

By finite frame logic we mean the logic that is given by frame logic with certain additional constraints on the theory-net $\mathbf{N}$ and the language $L_F(\mathbf{N})$:

Explanation 8.2 (Finite frame logic) The relations $\vdash_F$ and $\models_F$ defined for a theory-net $\mathbf{N}$ and a language $L_F(\mathbf{N})$ form a logic of finite frame logic if and only if

(1) the set of all applications in $\mathbf{N}$ is finite.

[3]The notion of a P-consequence is relative to the set of models of a set of P-axioms. Hence, any P-axiom is a P-consequence of the empty set. Our consequence relation among theoretical sentences (Definition 5.5) is inspired by this notion of a P-consequence.

[4]See Ebbinghaus and Flum [50, Chap. 3.5], for the proof that finite first-order logic cannot be axiomatised.

(2) For any application α, the domain D of α has finite cardinality.
(3) Any object of any domain of an application is named by a constant.

While these constraints might not be considered very elegant from a purely logical point of view, they are quite natural when it comes to investigating means of achieving tractability of belief changes. If the naming assumption is violated, there are frame-logical consequences that cannot be captured by syntactic derivations. For, the postulates **P**4.1(**T**) and **P**4.2(**T**) apply to tuples of objects independently of whether the objects in the tuples are named or not, while the inference rules Link Exploitation 1 and 2 are only applicable if the respective objects are named. Assuming these constraints, we can show that finite frame logic is complete for conclusions that are free of quantifiers:

Theorem 8.2 (Completeness) *Let Γ be a set of $L_F(\mathbf{N})$ sentences. ϕ is a quantifier-free sentence of $L_F(\mathbf{N})$. Suppose $\vdash_F$ and $\models_F$ form a system of finite frame logic in the sense of Explanation* 8.2. *If $\Gamma \models_F \phi$, then $\Gamma \vdash_F \phi$.*

The proof of completeness (to be found in the Appendix) makes essential use of the constraints of finite frame logic. Moreover, as regards the inference rules of intramodular reasoning, this proof builds upon completeness of first-order logic in an obvious way. Notably, when seeking an explanation of how quotidian reasoning is computationally feasible, we may well restrict our analysis to quantifier-free sentences.

8.2.5 Tractability

Let us now further assume that the modular units of reasoning are bounded by a constant. We express this assumption in terms of the cardinality of the set $V(\alpha)$, the set of descriptive symbols of an application α. Recall that, for any application α, there is a specific set $V(\alpha)$ of descriptive symbols. We say that the modular units of reasoning are bounded by a constant if and only if there is an upper bound on $|V(\alpha)|$ that applies to all α in **N**. On this condition, we can show that finite frame logic is tractable:

Theorem 8.3 *Let Γ be a finite set of $L_F(\mathbf{N})$ sentences. Suppose $\vdash_F$ and $\models_F$ form a system of finite frame logic. Further, suppose there is a constant c_1 such that, for all applications α in* **N**, *$|V(\alpha)| \leq c_1$. Then, the problem of determining whether Γ is frame-logically satisfiable has* **P***-time complexity.*

We prove this theorem in a manner that works out the crucial parts in detail, while an easily provable lemma is taken for granted. Let us recall some symbols already introduced in the previous chapter. $\mathbf{A}(\alpha)$ is the set of admissible interpretations of the application α. $H(\mathbf{N})$ designates the set of applications in **N**. Further, let m designate $|\Gamma|$ and n the number of applications in **N**. Since, for any application α in **N**, every object is designated by a constant, we can interpret $L(V(\alpha))$ by an assignment of truth values to the atoms of $L(V(\alpha))$. So, we take $\mathbf{A}(\alpha)$ to be a set of truth-value interpretations.

We can test for satisfiability of the set Γ using the following procedures. First, determine, for any application α in $\mathbf{N}$, the set of interpretations that satisfy all sentences $\phi|_\alpha \in \Gamma$ as well as all sentences $\phi|_A \in \Gamma$ such that $\alpha \in A$. If, for some application α, this set is empty, Γ is not satisfiable, and we are done. Then, we apply the semantic constraints of link satisfaction, as defined by the postulates **P**4.1(**T**) and **P**4.2(**T**) (Sect. 8.1). If, as a result of this application, there is a pair (α, β) of applications such that $\mathbf{A}(\alpha)$ and $\mathbf{A}(\beta)$ violate some link, Γ is not satisfiable. Otherwise, it is satisfiable. This suggests the following algorithm:

Algorithm 8.1 (Satisfiability of a set Γ of $L_F(\mathbf{N})$ sentences)

```
                    // Part A: determining intramodular inferences

for all α ∈ H(N):
   A(α) := {x | x ⊨ Γ_α}           // Γ_α is the set of sentences φ|_α ∈ Γ united with
                                    // the set of sentences φ|_A ∈ Γ such that α ∈ A
   if A(α) = ∅:
      output: not satisfiable;
      return;

                    // Part B: determining transmodular inferences

repeat
   determination := false          // determination is a flag that indicates if some
                                   // link has been used to determine some set A(α)
   for all α ∈ H(N):               // i.e., the following block of commands needs
                                   // to be carried out for all members of H(N)
      for all β ∈ H(N) \ {α}:
         for all ground literals l ∈ L(V(α)) ∩ L(V(β)):
            for all x ∈ A(α):
               if x ⊭ l and for all x ∈ A(β), x ⊨ l:
                  A(α) := A(α) \ {x};
                  determination := true;
                  if A(α) = ∅:
                     output: not satisfiable;
                     return;
until not determination;
                                   // i.e., until no set A(α) gets determined any more
output: satisfiable.
```

It is almost obvious and easy to prove explicitly that this algorithm determines whether or not there is an $L(\mathbf{N})$ interpretation such that all members of Γ are true. That is, the output is *satisfiable* if and only if the set Γ is in fact satisfiable, according to the semantics of frame logic. Using this claim as a lemma, we can prove the theorem by showing that the maximal number of computation steps is bounded by a polynomial in the parameters n and m. As indicated above, the algorithm consists of two parts. Part A determines, for all $\alpha \in H(\mathbf{N})$, the set $\mathbf{A}(\alpha)$ without considering links. We have described this part as intramodular reasoning. Part B, by contrast, determines the semantic consequences of link satisfaction for the sets $\mathbf{A}(\alpha)$ for all $\alpha \in H(\mathbf{N})$. We have described this part as transmodular reasoning.

Since we assumed an upper bound on $|V(\alpha)|$, it takes a constant number of computation steps to determine which interpretations x of $L(V(\alpha))$ satisfy a sentence $\phi|_\alpha$ or a sentence $\phi|_A$, where $\alpha \in A$. Hence,

$$Comp(intra) = O(n \cdot m)$$

where *Comp(intra)* stands for the computational complexity of the first part of the algorithm.

As for the second part, we need to determine the complexity of determining, for a given pair (α, β) $(\alpha, \beta \in H(\mathbf{N}))$ the literals l such that there is $x \in \mathbf{A}(\alpha)$, $x \not\models l$, while for all $x \in \mathbf{A}(\beta)$, $x \models l$. Since $|V(\alpha)|$ is bounded by a constant and since every object in α's domain D of interpretation is designated by a constant in $V(\alpha)$, $|\mathbf{A}(\alpha)|$ as well as the number of ground literals in $L(V(\alpha)) \cap L(V(\alpha))$ are bounded by respective constants. Hence, determining, for a given pair (α, β), the literals l such that there is $x \in \mathbf{A}(\alpha)$, $x \not\models l$ – while for all $x \in \mathbf{A}(\beta)$, $x \models l$ – takes at most a constant number of computation steps. Obviously, this determination of literals is carried out for each pair (α, β) such that $\alpha, \beta \in H(\mathbf{N})$. Hence, (i) the complexity of running an instance of the repeat-until loop in Algorithm 8.1 is $O(n^2)$.

Note, however, that there is upper bound on the number of repeat-until loops necessary for the termination of the algorithm. For there to be such a bound, it is in turn crucial that $|V(\alpha)|$ is bounded by a constant c_1. Because of the latter bound, the number of truth-value interpretations of $L(V(\alpha))$ is bounded by a constant, say c_3. Hence, the process of detecting ground literals l that are not verified by all $x \in \mathbf{A}(\alpha)$, but verified by all $x \in A(\beta)$ must come to an end, after a certain number of instances of the repeat-until loop. For, this is a monotonic process, in the sense that at least one set $\mathbf{A}(\alpha)$ becomes smaller after an instance of the repeat-until loop, except for the last instance where the determination flag remains false. Even in the worst case, (ii) the algorithm takes less than $n \cdot c_3$ instances of the repeat-until loop to terminate. (i) and (ii) imply

$$Comp(trans) = O(n^3)$$

where *Comp(trans)* stands for the second part of the algorithm, which determines the semantic consequences of transmodular reasoning. Now, $Comp(intra) = O(n \cdot m)$ and $Comp(trans) = O(n^3)$ imply that the complexity of the present algorithm is $O(n^3)$. Hence, the problem of determining whether or not a set of $L_F(\mathbf{N})$ sentences is satisfiable is **P**-time complex on condition that $|V(\alpha)|$ is bounded by a constant. This concludes the proof of Theorem 8.3.

Thus, we have shown that the satisfiability problem is fixed-parameter tractable for finite frame logic. The crucial parameter is the syntactic size of the modular units of reasoning. Without assuming an upper bound on $|V(\alpha)|$, frame logic is not tractable. The result is significant for cognitive science only if $|V(\alpha)|$ is "sufficiently small". While it is notoriously difficult to make the latter notion precise, we have argued in Sect. 7.3.4 that this is a reasonable assumption. To be precise, we have argued there that the number of interpretations of $L(V(\alpha))$ is bounded by a small constant or that the models of the substantial law can be determined efficiently for all applications α, or both.

A few notes relating the present tractability result to the tractability result of Theorem 7.1 (obtained for the TMS) are in order. The TMS algorithm is not simple at all, but aims to optimise the determination of semantic consequences as much as possible. By contrast, the above algorithm for the test of satisfiability is simple, but not very sophisticated. This algorithm involves a number of redundant tests. However, Algorithm 8.1 is sophisticated enough to show that the problem in question is tractable.

Note, furthermore, that the above algorithm and the TMS algorithm solve different, albeit related problems. The TMS determines how the sets $\mathbf{A}(\alpha)$ and the acceptance of applications change when receiving new epistemic input in the form of a new application. This determination builds upon previously drawn semantic consequences. The above algorithm, by contrast, starts determining semantic consequences from scratch. It tests for the satisfiability of a set of sentences, but does not determine nonmonotonic belief changes.

8.2.6 Composition

In frame logic, certain first-order derivations are not available, even though first-order logic is available within a single frame instance. Otherwise, frame logic would not be tractable. For example, from

$$\{p \vee q|_\alpha, p \to s|_\beta, q \to s|_\gamma\}$$

we cannot infer $s|_\beta$. These inferences, however, would have been valid on the semantics of $\mathbf{AE}(\mathbf{T})$, with the original $P4(\mathbf{T})$ in place of its descendants $P4.1(\mathbf{T})$ and $P4.2(\mathbf{T})$. In light of such cases of relative incompleteness – relative to full first-order reasoning without a modular division of the language into frames – one must wonder whether it makes sense to introduce an inference rule *Composition*:

$$\frac{\phi|_\alpha \qquad \psi|_\beta}{\phi|_\gamma \qquad \psi|_\gamma} \qquad [\text{where } \gamma = C(\alpha, \beta)]. \qquad \text{(Composition)}$$

$\gamma = C(\alpha, \beta)$ simply says that γ is a composition out of α and β. The operation of forming composite applications has been described in Sect. 8.2.6. On the one hand, such a logic would allow us to overcome the inferential limitations of frame logic in the sense that frame logic with Composition would be as inferentially powerful as first-order logic. For, Composition would allow us to put any assertion into an "overarching" frame that encompasses all assertions. Within this overarching frame, any first-order proof could be carried out. On the other hand, frame logic with the inference rule Composition would not be tractable any more. In other words, frame logic with Composition collapses into standard first-order logic. Such a logic is neither decidable nor tractable.

We may nonetheless allow compositions of intended applications but consider such operations to be extra-logical. For, a composition changes the division of the global domain into subdomains and may, therefore, change the frame-logical consequences of a given set Γ. Thereby, the extension of $\models_F$ would change. Hence, we do not introduce Composition as a valid inference rule of frame logic. Composition of applications is an extra-logical operation, in the sense that it cannot be accomplished within the inferential calculus.

Note, finally, that the tractability result of Theorem 8.3 applies without qualifications to sets of $L_F(\mathbf{N})$ sentences even if $\mathbf{N}$ has composite applications. For, no assumption has been made in Theorem 8.3 and its proof as to whether there are composite applications. As long as the syntactic size of the applications is bounded by a constant, frame logic is (fixed-parameter) tractable.

8.3 Resolution Frame Logic

Having devised a natural deduction system of frame logic, we shall now develop a variant of this logic using the resolution calculus. What are the benefits of using resolution for frame logic? The main benefit is simplicity. More specifically, the resolution framework allows us to implement the distinction between intra- and transmodular reasoning in a relatively simple and elegant way. Moreover, the operations of truth maintenance can be simplified using resolution frame logic instead of the postulates $\mathbf{P}_F(\mathbf{N})$ of the structuralist framework. Finally, proofs of tractability, soundness and completeness can be given in a simpler fashion when working with the resolution framework instead of a natural deduction system and the postulates $\mathbf{P}_F(\mathbf{N})$. Yet, the structuralist framework remains in the background of resolution frame logic. Another benefit of using the resolution calculus is that the problem of weak determinations (Sect. 7.4.6) does not arise in the first place.

8.3.1 The Resolution Principle

The resolution calculus is widely used for automated theorem proving and knowledge representation. It goes back to Davis and Putnam [44], and Robinson [96].[5]

[5]For a thorough introduction to the resolution calculus for propositional and first-order logic, the reader is referred to Schöning [102].

The calculus is based on a single inference rule, called *Resolution*. Let $\alpha_1 \vee \ldots \vee \alpha_{i-1} \vee \gamma \vee \alpha_i \vee \ldots \vee \alpha_n$ and $\beta_1 \vee \ldots \vee \beta_{j-1} \vee \neg\gamma \vee \beta_j \vee \ldots \vee \beta_k$ be disjunctions of ground literals. (A ground literal is a literal that has no occurrences of a variable.) The resolution inference rule goes as follows:

$$\frac{\begin{array}{c}\alpha_1 \vee \ldots \vee \alpha_{i-1} \vee \gamma \vee \alpha_i \vee \ldots \vee \alpha_n,\\ \beta_1 \vee \ldots \vee \beta_{j-1} \vee \neg\gamma \vee \beta_j \vee \ldots \vee \beta_k\end{array}}{\alpha_1 \vee \ldots \vee \alpha_n \vee \beta_1 \vee \ldots \vee \beta_k}.$$

To give a simple example of an application of this inference rule:

$$\frac{p \vee q, \quad \neg p \vee r}{q \vee r}.$$

Recall that a literal is an atomic formula or the negation of such a formula. If l and l' are literals such that $Cn(l) = Cn(\neg l')$, then l is called the *complementary literal* of l', and vice versa. The literal complementary to l is designated by $\overline{l}$. For example, $\overline{p}$ is $\neg p$, while $\overline{\neg p}$ is p. Obviously, when resolving a literal l with its complementary literal $\overline{l}$, we obtain a contradiction.

The resolution inference rule is applicable only to pairs of formulas that are disjunctions of ground literals or ground literals. It is common to represent a disjunction of literals by the set of its disjuncts. Such a set is called a *clause*. $\Box$ designates the empty clause. Since we can derive this clause only by resolving a ground literal with its complementary literal, deriving the empty clause amounts to deriving a contradiction.

Using clauses, we can express the resolution inference rule in a more concise way. Let C and C' be ground clauses, i.e., sets of ground literals. Further, let l be a ground literal. Then, resolving C with C' goes as follows:

$$\frac{C \quad C'}{C - l \cup C' - \overline{l}} \qquad \text{(Resolution)}$$

where $C - l$ is shorthand for $C \setminus \{l\}$.[6]

For full propositional reasoning in the resolution calculus, we need to transform premises into conjunctive normal form. The propositional resolution calculus is sound and refutation complete with respect to the semantics of classical propositional logic[7]:

Theorem 8.4 (Resolution theorem (of propositional logic)) *A set Γ of ground clauses is classically unsatisfiable if and only if $\Gamma \vdash_R \Box$, i.e., if and only if a contradiction can be derived from Γ using Resolution.*

---

[6]These notational conventions follow standard expositions of the resolution calculus (see, e.g., Schöning [102]).

[7]For the proof of this theorem, see See Schöning [102, Chap. 1.5].

Using $\Gamma \models \alpha$ if and only if $\Gamma \cup \{\neg\alpha\}$ is not satisfiable, propositional logical entailment can be tested using the resolution calculus. The inference rule Resolution is also referred to as *ground resolution*.

To capture full first-order reasoning, premises need to be transformed into Skolem normal form [102, Chap. 2.5]. Moreover, there is a generalisation of the ground resolution inference rule that is applicable to disjunctions of non-ground literals, using a substitution method called *unification*. For the present investigation, we confine ourselves to ground resolution. When devising resolution frame logic in the next section, we shall exploit the resolution theorem of propositional logic.

8.3.2 Frame Logic with Resolution

The core principles of frame logic can be implemented into the (ground) resolution calculus in a surprisingly simple fashion. Recall the two principles of frame logic: within an application, full first-order logic is available. Among applications, however, only ground literals can be communicated. In what follows, only classical propositional reasoning is available within an application.

Let us devise the syntax and semantics of resolution frame logic in a manner analogous to natural deduction frame logic, as expounded in the previous section. Resolution frame logic centres on sets of ground literals that are meant to represent disjunctions of ground literals. Recall that such sets are called *ground clauses*. We, therefore, designate clauses by symbols of the form C_α, C'_α, C_β, etc. The index of these upper case letters stands for an application. This is the alphabet of $L_{RF}(\mathbf{N})$, the resolution frame language associated with $\mathbf{N}$:

Definition 8.7 (Alphabet of $L_{RF}(\mathbf{N})$) Let $\mathbf{N}$ be a theory-net. The alphabet of $L_{RF}(\mathbf{N})$ consists of

(1) the logical symbols $\neg$, $-$, and $\Box$
(2) a set N_a of names for applications of theory-elements
(3) for every $\alpha \in N_a$, the set $V(\alpha)$ of descriptive symbols
(4) upper case letters, such as C, C', C'', etc.
(5) schematic letters for literals and complements of literals.

Well-formed formulas of $L_{RF}(\mathbf{N})$ are built as follows:

Definition 8.8 (Well-formed formula of $L_{RF}(\mathbf{N})$) Let $L_{RF}(\mathbf{N})$ be the resolution frame language of the theory-net $\mathbf{N}$.

(1) If l is a ground literal of $L(V(\alpha))$ and $\alpha \in N_a$, then l_α and $\bar{l}_\alpha$ are formulas of $L_{RF}(\mathbf{N})$.
(2) If $l_1, \ldots, l_n$ are ground literals of $L(V(\alpha))$, $\alpha \in N_a$ and $C = \{l_1, \ldots, l_n\}$, then C_α and $\neg C_\alpha$ are formulas of $L_{RF}(\mathbf{N})$.
(3) If C_α is an $L_{RF}(\mathbf{N})$ formula and l a literal of $L(V(\alpha))$, then $(C - l)_\alpha$ is an $L_{RF}(\mathbf{N})$ formula.

(4) If C_α and C'_α are formulas of $L_{RF}(\mathbf{N})$, then $(C \cup C')_\alpha$ is an $L_{RF}(\mathbf{N})$ formula.
(5) Nothing else is a formula of $L_{RF}(\mathbf{N})$.

At the core of resolution frame logic, we have inference rules for intramodular and transmodular resolution:

$$\frac{C_\alpha \quad C'_\alpha}{(C - l \cup C' - \bar{l})_\alpha} \qquad \text{(Intramodular Resolution)}$$

$$\frac{C_\alpha \quad l_\beta}{(C - \bar{l})_\alpha}. \qquad \text{(Transmodular Resolution)}$$

Recall that upper case letters, such as C, C', etc. designate sets of literals. The intended meaning of $C = \{l_1, \ldots, l_n\}$, however, is to represent the disjunction $l_1 \vee \ldots \vee l_n$. So there is a mismatch between the intended meaning and the literal meaning of the symbols C, C', etc. But this mismatch greatly simplifies the exposition of the resolution calculus, which is why it is widely adopted in the knowledge representation literature.

The following inference rule may be described as *contradiction introduction*:

$$\frac{l_\alpha \quad \bar{l}_\beta}{\Box} \quad [\alpha = \beta \text{ or } \alpha \neq \beta]. \qquad \text{(Contradiction Introduction)}$$

Literals and disjunctions of literals are to be derived by means of an indirect proof:

$$\frac{\neg C_\alpha \vdash_s \Box}{C_\alpha} \qquad \text{(Negation Elimination 1)}$$

$$\frac{\bar{l}_\alpha \vdash_s \Box}{l_\alpha} \qquad \text{(Negation Elimination 2)}$$

$$\frac{C_\alpha}{l_\alpha} \quad [C = \{l\}] \qquad \text{(Literal Introduction 1)}$$

$$\frac{\neg C_\alpha}{l_\alpha} \quad [\bar{l} \in C] \qquad \text{(Literal Introduction 2)}$$

$$\frac{\neg C_\alpha}{\bar{l}_\alpha} \quad [l \in C]. \qquad \text{(Literal Introduction 3)}$$

The latter group of rules goes beyond the standard resolution calculus. The rules of this group allow us to prove literals and clauses from sets of clauses. This is necessary since we want to exploit the determinate literals of one application for the inferences of another application. Note that $\neg C$ stands for a negated disjunction of literals. Hence, if $C = \{l_1, \ldots, l_n\}$, then $\neg C$ has the intended meaning of $\neg l_1 \wedge \ldots \wedge \neg l_n$.

Note that we use symbols of the form C to designate a set of literals and symbols of the form C_α to designate a set of literals that is relativised to an application α. To be more precise, the set C_α of literals is relativised to the language $L(V(\alpha))$. Thus, the structuralist framework merely serves the purpose of dividing a given formal language into modular units. Unlike in natural deduction frame logic, there is no inference rule that may be described as *law application*. Consequently, substantial laws are not explicitly introduced. But implicitly it is assumed that instances of such laws are represented by corresponding clauses or corresponding sets of clauses. This implicit assumption, however, is not needed for the core formalism of resolution frame logic. The definition of a derivation in resolution frame logic is straightforward[8]:

Definition 8.9 ($\Gamma \vdash_{RF} \phi$) Let Γ be a set of $L_{RF}(\mathbf{N})$ sentences and let ϕ be such a sentence. $\Gamma \vdash_{RF} \phi$ if and only if there is a tree of $L_{RF}(\mathbf{N})$ sentences such that

(1) the topmost sentences are either in Γ or premises of a completed subproof. A subproof is complete if and only if it has been used as a justification of an application of an inference rule.
(2) The bottommost sentence is ϕ.
(3) Every sentence in the tree, except ϕ, is a premise of an application of an inference rule (of resolution frame logic) such that the conclusion of this application stands directly below that sentence.

Let us move on to the semantics of resolution frame logic. Since we have confined our language to ground literals, we can spell out the semantics using truth-value interpretations. Let $\mathcal{L}_+(N)$ be the set of literals l_α such that there is $\alpha \in N_a$ and l is a positive ground literal of $L(V(\alpha))$. Then, we can say that an $L_{RF}(\mathbf{N})$ interpretation $\mathcal{V}(\mathbf{N})$ is a mapping of the set of positive ground literals of $\mathbf{N}$ onto the set of the classical truth values. In symbols, $\mathcal{V}(\mathbf{N}) : \mathcal{L}_+(\mathbf{N}) \to \{T, F\}$.

When does an interpretation $\mathcal{V}(\mathbf{N})$ satisfy a set Γ of sentences of this language? The notion of satisfaction, in turn, is guided by the distinction between intramodular and transmodular reasoning. More specifically, satisfaction of a set of clauses is understood as satisfaction of a corresponding set of disjunctions of literals, where negation and disjunction have their classical meanings. In addition to this intramodular condition, an interpretation $\mathcal{V}(\mathbf{N})$ must satisfy a transmodular condition: if a literal l_α is determined to be true, then l_β must be determined to be true as well. To understand the wording of this constraint, recall that l designates a literal and l_α a relativised literal. l is a symbol of a standard first-order language, while l_α is a symbol of the corresponding modularised language, termed *frame language*.

To express the intramodular and the transmodular semantic constraints in a precise way, we relativise the interpretation $\mathcal{V}(\mathbf{N})$ to applications. Let $\mathcal{V}(\mathbf{N})_\alpha$ be the restriction of the interpretation $\mathcal{V}(\mathbf{N})$ to the language $L(V(\alpha))$. That is, $\mathcal{V}(\mathbf{N})_\alpha$ is

[8]The wording of the following definition draws on Definition 9.2 in Zach [122].

a mapping of the set of positive $L(V(\alpha))$ ground literals to the set of classical truth values. Likewise, Γ_α is the set of sentences ϕ such that $\phi_\alpha \in \Gamma$. In more formal terms, $\Gamma_\alpha = \{\phi \mid \phi_\alpha \in \Gamma\}$.

There are at least two ways to express the transmodular semantic constraint. First, by requiring that pairs of interpretations of the sublanguages $L(V(\alpha))$, $L(V(\beta))$, etc. satisfy a certain constraint that is analogous to the semantic constraints expressed by the postulates **P**4.1(**T**) and **P**4.2(**T**). Second, we can use the fixed point of an operation that exploits links among applications. The second option seems to allow for a simpler definition, which is why we pursue this option here:

Definition 8.10 (Γ^*)
Let Γ be a set of $L_{RF}(\mathbf{N})$ sentences. Γ^* is defined as follows:

$$\Gamma^0 = \Gamma$$

$$\Gamma^{n+1} = \Gamma^n \cup \{l_\beta \mid \text{there are } \alpha \text{ and } \beta \text{ s.t. } l \in L(V(\beta)) \text{ and } \Gamma^n_\alpha \models l\}$$

$$\Gamma^* = \Gamma^{n+1} \text{ s.t. } \Gamma^{n+1} = \Gamma^n, \text{ and } \Gamma^n \neq \Gamma^{n-1} \text{ or } n = 0.$$

Γ^* is the semantic inferential closure of an inference rule that exploits links between two modular units of reasoning. Such modular units are defined by sublanguages of the type $L(V(\alpha))$, $L(V(\beta))$, etc. Note that the definition of Γ^* uses $\models$ in the sense of classical logic. In the context of $\models$, clauses are taken to represent corresponding disjunctions of literals. Likewise, a negated clause represents the negation of a disjunction of literals. So, symbols for clauses are used heteronymously: in certain contexts they have their literal meaning (when they designate actual sets of literals), in others their intended meaning (when they are understood as actual disjunctions of literals). This may be confusing, but allows for a simple and intuitive notation. We define satisfaction of an $L_F(\mathbf{N})$ interpretation in terms of the satisfaction of Γ^*:

Definition 8.11 ($\mathcal{V}(\mathbf{N}) \models_{RF} \Gamma$) Let $\mathcal{V}(\mathbf{N})$ be an interpretation of a frame language $L_{RF}(\mathbf{N})$. Γ is a set of $L_{RF}(\mathbf{N})$ sentences. $\mathcal{V}(\mathbf{N}) \models_{RF} \Gamma$ if and only if, for all $\alpha \in N_a$, $\mathcal{V}(\mathbf{N})_\alpha \models \Gamma^*_\alpha$.

The relation of logical consequence can be defined in the standard way:

Definition 8.12 ($\Gamma \models_{RF} \phi$) Let Γ be a set of $L_{RF}(\mathbf{N})$ sentences and ϕ be such a sentence. $\Gamma \models_{RF} \phi$ if and only if, for all interpretations $\mathcal{V}(\mathbf{N})$, if $\mathcal{V}(\mathbf{N}) \models_{RF} \Gamma$, then $\mathcal{V}(\mathbf{N}) \models_{RF} \phi$.

Notice that we use the classical relations of satisfaction and logical consequence within the definition of further relations of satisfaction and logical consequence. No circularity arises from this. The relations of satisfaction defined, respectively, for the languages $L(\mathbf{N})$ and $L_F(\mathbf{N})$ rest on classical logic in a similar way.

8.3.3 Soundness, Completeness and Tractability

We establish soundness and completeness of resolution frame logic:

Theorem 8.5 (Soundness) *Let* Γ *be a set of* $L_{RF}(\mathbf{N})$ *sentences and* ϕ *be such a sentence. If* $\Gamma \vdash_{RF} \phi$*, then* $\Gamma \models_{RF} \phi$.

Theorem 8.6 (Completeness) *Let* Γ *be a set of* $L_{RF}(\mathbf{N})$ *sentences and* ϕ *be such a sentence. If* $\Gamma \models_{RF} \phi$*, then* $\Gamma \vdash_{RF} \phi$.

The proofs for these two theorems are given in the Appendix. The proof of completeness rests on the resolution theorem of propositional logic (Theorem 8.4). Since resolution frame logic is less expressive than natural deduction frame logic, the relations $\models_{RF}$ and $\models_F$ are not equivalent. Not surprisingly, the relations $\vdash_{RF}$ and $\vdash_F$ are not equivalent either. However, set aside the peculiar syntactic representations of disjunctions of literals in the resolution framework, it is easy to show that the relation $\models_{RF}$ is a proper subset of $\models_F$. Likewise, $\vdash_{RF}$ is a proper subset of $\vdash_F$.

Tractability is the key motivation for devising frame-logical systems of reasoning in the first place. Resolution frame logic is fixed-parameter tractable, just as natural deduction frame logic is:

Theorem 8.7 *Let* Γ *be a finite set of* $L_{RF}(\mathbf{N})$ *sentences. Further, suppose there is a constant* c_1 *such that, for all applications* α *in* $\mathbf{N}$, $|V(\alpha)| \leq c_1$. *The problem of determining whether* Γ *is satisfiable – in the sense of resolution frame logic – has* **P***-time complexity.*

There are two ways to prove this theorem. First, we can translate the semantic operations of Algorithm 8.1 into syntactic operations on clauses and sets of clauses such that the inference rules of resolution frame logic are implemented. Then, we show that the resulting algorithm has **P**-time complexity. Using soundness and completeness of resolution frame logic, we can infer from this Theorem 8.7. Second, we can show that the semantic operations of Algorithm 8.1 capture the semantics of resolution frame logic correctly and completely. Once this has been done, tractability of resolution frame logic follows from Algorithm 8.1 having **P**-time complexity. The latter result has been established when proving tractability of natural deduction frame logic (Theorem 8.3). We follow the second way of proving Theorem 8.7 in the Appendix.

8.4 Tractable Belief Revision

Natural deduction and resolution frame logic are tractable monotonic logics. We devised these logics with the intent to achieve a tractable account of belief changes. Recall, however, that a tractable monotonic logic alone does not suffice to achieve tractability of belief revision (Sect. 7.1.2). This is why we need to adjust also the belief revision scheme. In the next section, we shall modify the belief revision scheme of preferred subtheories in light of the complexity result for the TMS in the previous chapter. Further, we shall explain and adopt a syntactic understanding of applications such that any instance of a substantial law forms a distinct application. These two further changes yield an account of belief revision that is tractable for resolution frame logic.

8.4.1 Avoiding Irrelevant Determinations

In this section, we adopt and explain a syntactic understanding of applications and a constraint on the logical form of substantial laws. These changes are motivated by the problem of irrelevant determinations, described in Sect. 7.4.5. This problem arises from the general problem of determining which applications are relevant for an inconsistency detected between an application α and an application β. More precisely, the problem is to determine which of the applications that – directly or indirectly – determine α are decisive for an inconsistency between α and β.

The inference rule Resolution applies directly only to pairs of formulas that are disjunctions of literals or literals. This suggests a restriction on the logical form of substantial laws to universally quantified implications. More precisely, we require any substantial law of any theory-element to have the following logical form:

$$\forall x_1 \ldots \forall x_n (\phi_0 \wedge \ldots \wedge \phi_k \rightarrow \phi_{k+1}) \qquad (n \geq 1, k \geq 0) \qquad (8.1)$$

where $\phi_0, \ldots, \phi_{k+1}$ are atomic formulas. Virtually all strict and non-strict laws can be represented using laws of this form. k is admitted to be zero to allow for universally quantified atomic formulas. Recall that a biimplication is logically equivalent to a set of two implications so that the former can be represented by the latter. An implication of the form

$$\phi_1 \rightarrow \phi_2 \wedge \phi_3$$

can be represented by the implications

$$\phi_1 \rightarrow \phi_2 \qquad \text{and} \qquad \phi_1 \rightarrow \phi_3.$$

The present restriction on the logical form of substantial laws excludes that an existential quantifier occurs in a substantial law. We can lift this restriction using the method of transforming a formula into Skolem normal form.[9] For simplicity, we accept the loss of expressivity coming with the presently proposed restriction.

[9]This method requires the introduction of function symbols, which we avoided so far.

To address the problem of irrelevant determinations, we need to go one step further. We need to adopt a syntactic understanding of applications such that any instance of a substantial law forms a distinct application. Why so? Take the following example of a universally quantified implication:

$$\forall x \forall y (R_1(x, y) \rightarrow R_2(x, y)). \tag{$\lambda(\mathbf{T})$}$$

This law has applications in domains with at least two objects. Suppose the two objects are designated by a and b, and a and b are descriptive symbols of $L(V(\alpha))$. Further, suppose α is an application of $\mathbf{T}$, and the substantial law is given by $\lambda(\mathbf{T})$. Let the domain of α be given by $D(\alpha)$, where $|D(\alpha)|=2$. Then, $\lambda(\mathbf{T})$ has two instances in $D(\alpha)$:

$$R_1(a, b) \rightarrow R_2(a, b) \tag{ϕ_1}$$

$$R_1(b, a) \rightarrow R_2(b, a). \tag{ϕ_2}$$

Suppose an application β_1 directly determines $R_1(a, b)$ to be true, while an application γ directly determines $R_2(a, b)$ to be false. Further, β_2 directly determines $R_1(b, a)$ to be true, while there is no application that directly determines $R_2(b, a)$. All applications in play have the same epistemic rank. In this situation, our TMS doubts α, β_1, β_2 and γ. But β_2 is not relevant for the inconsistency since the inconsistency hinges only on ϕ_1 but not on ϕ_2. To be more precise, β_2 is a member of all preferred subtheories of $\Theta(\mathbf{N})$, provided there are no applications in $\mathbf{N}$ other than α, β_1, β_2 and γ. (Recall that β_2 designates a sentence of the form $c \in \mathbf{I}(\mathbf{T})$ so that it makes sense to say of β_2 that it is a member of a preferred subtheory of $\Theta(\mathbf{N})$.)

The present example suggests a relatively simple solution of the general problem: instead of a semantic understanding of applications, we adopt a syntactic understanding such that every single instance of the substantial law of $\mathbf{T}$ forms a separate application. Applied to the example, this means that ϕ_1 and ϕ_2 are considered distinct applications. This leads to a one-to-one correspondence between applications and instances of substantial laws. Then, we can always determine which applications are relevant for an inconsistency observed between two applications. This will be shown when proving Proposition 8.3 below.

In sum, we adopt a constraint on the logical form of substantial laws such that every substantial law must be a universally quantified implication of the form (8.1). Further, applications are given by single instances of substantial laws such that there is a one-to-one correspondence between applications and such instances. Such a syntactic understanding of applications is consistent with the syntax and semantics of resolution frame logic. We merely need to be prepared for applications α_1 and α_2 such that $V(\alpha_1) = V(\alpha_2)$. Neither the semantics nor the syntax of resolution frame logic require that $V(\alpha_1) \neq V(\alpha_2)$. The transition from a semantic to a syntactic understanding of applications lines up very nicely with the transition from a semantic TMS to a proof-theoretic one that implements the calculus of resolution frame logic. This transition will be worked out in what follows.

8.4.2 Weak Determinations Resolved

Recall the problem of weak determinations from Sect. 7.4.6. It arises from semantic determinations between formulas of the following types:

$$q \qquad (\alpha_1)$$

$$p \to q \qquad (\alpha_2)$$

$$\neg q. \qquad (\alpha_3)$$

Suppose α_1 determines α_2, in the sense that the dynamic set $\mathbf{A}(\alpha_2)$ is diminished as a consequence of testing link satisfaction in the TMS. Then, q will be determined true in α_2. Further, suppose the pair (α_2, α_3) is tested for link satisfaction. Then, α_2 is found to not satisfy all links to α_3. So, α_2 may be rejected or doubted, depending on the epistemic ranks of α_1, α_2 and α_3. But this is not correct since α_2 is a member of all preferred subtheories of the corresponding theory $\Theta(\mathbf{N})$, provided there are no further applications in play.

Fortunately, this type of problem does not arise in the resolution calculus. For this to be seen, let us rewrite the above propositional formulas using disjunctions of literals and literals:

$$q \qquad (\alpha'_1)$$

$$\neg p \vee q \qquad (\alpha'_2)$$

$$\neg q. \qquad (\alpha'_3)$$

Obviously, α'_1 cannot be resolved with α'_2. Hence, α'_1 does not determine α'_2 when determinations among applications are implemented using the resolution calculus.

8.4.3 TM Belief Revision

In this section, we devise a belief revision scheme that mirrors the working of the TMS. This is why we call it *TM belief revision scheme*. The key idea is to define, in a recursive fashion, the division of applications into successful, doubted and unsuccessful ones. The corresponding operations of acceptance, doubt and rejection are motivated by the operations of the TMS, but may also be understood as technical terms defined within the below belief revision scheme. The TM belief revision scheme is thus motivated by the TMS, but defined independently of that system. This enables us to show tractability of the belief revision scheme – for resolution frame logic – without relying on the tractability result about the TMS given by Theorem 7.1.

Let us recall the intended meanings of the operations of acceptance, doubt and rejection. The notion of a successful application is intended to capture those applications that are a member of all preferred subtheories of the respective default theory $\Theta(\mathbf{N})$. The notion of a doubted application is meant to recognize the applications that are in some but not all preferred subtheories of $\Theta(\mathbf{N})$. The set of rejected applications, finally, is intended to collect the applications that are not a member of any preferred subtheory of $\Theta(\mathbf{N})$. Such applications are also called *unsuccessful*.

Let $\Theta = \langle T_1, \ldots, T_n \rangle$ be a default theory that serves as a ranked belief base. Further, let $S = \langle \alpha_1, \ldots, \alpha_k \rangle$ be a sequence of revision formulas that we receive as epistemic input, starting from an empty belief base $\Theta_0 = \langle \emptyset, \ldots, \emptyset \rangle$. The union of the members of S equals the union of $T_1, \ldots, T_n$. In symbols, $\{\alpha_1\} \cup \ldots \cup \{\alpha_k\} = T_1 \cup \ldots \cup T_n$. We assume that the order of the sequence S conforms to the epistemic ranking Θ. That is, if $i < j$ $(1 \leq i, j \leq n)$, then α_i has at least the epistemic rank of α_j. B_i designates the set of explicit beliefs in Θ at stage i of the revision process. That is, B_i contains those formulas of the sequence S that are believed after the empty belief base has been revised by $\alpha_1, \ldots, \alpha_i$. Further, let D_i designate the set of doubted beliefs in Θ at stage i of the revision process. R_i, finally, designates the set of rejected beliefs in Θ at stage i of the revision process. For any set A of formulas of Θ, $(A)_{<\alpha}$ designates the formulas of A that are epistemically superior to α, while $(A)_{=\alpha}$ designates the formulas of A that have the same epistemic rank as α. $(A)_{\leq\alpha}$ has its obvious meaning.

Even though the belief revision scheme is intended to be used for frame logic, we define it in a general way. $\vdash$, $\models$ and Cn, thus, stand for whatever logic chosen to govern the beliefs of the revision scheme. We define the division into actually believed, doubted and rejected beliefs in a recursive fashion:

Definition 8.13 (TM belief revision) Let $\Theta = \langle T_1, \ldots, T_n \rangle$ be a default theory and $S = \langle \alpha_1, \ldots, \alpha_k \rangle$ a sequence of revision formulas that respects the epistemic ranking of Θ. Let $\mathbf{S}$ be the set of such sequences.

$$B_1 = \{\alpha_1\} \text{ iff } \alpha_1 \nvdash \bot$$

$$D_1 = \emptyset$$

$$R_1 = \{\alpha_1\} \text{ iff } \alpha_1 \vdash \bot$$

$$b(\alpha_{i+1}) \text{ iff, for all } \delta \in D_i,\ B_i \cup \{\alpha_{i+1}, \delta\} \nvdash \bot$$

$$r(\alpha_{i+1}) \text{ iff } (B_i)_{<\alpha_{i+1}}, \alpha_{i+1} \vdash \bot$$

$$d(\alpha_{i+1}) \text{ iff } (B_i)_{\leq\alpha_{i+1}}, \alpha_{i+1} \vdash \bot \text{ and not } r(\alpha_{i+1})$$

$$d'(\alpha_{i+1}) \text{ iff not } b(\alpha_{i+1}) \text{ and not } r(\alpha_{i+1}) \text{ and not } d(\alpha_{i+1})$$

$$\mathbf{\Delta}_{i+1} = \{\Delta \mid \Delta \subseteq (B_i)_{=\alpha_{i+1}} \text{ and } (B_i)_{<\alpha_{i+1}} \cup \Delta \cup \{\alpha_{i+1}\} \text{ is minimally inconsistent in } \Delta \cup \{\alpha_{i+1}\}\}$$

$$\mathbf{\Delta}'_{i+1} = \{\Delta \mid \Delta \subseteq (B_i)_{=\alpha_{i+1}} \text{ and there is } \delta \in D_i \text{ s.t. } (B_i)_{<\alpha_{i+1}} \cup \Delta \cup \{\delta, \alpha_{i+1}\} \text{ is minimally inconsistent in } \Delta \cup \{\delta, \alpha_{i+1}\}\}$$

$$B_{i+1} = \begin{cases} B_i \cup \{\alpha_{i+1}\} & \text{if } b(\alpha_{i+1}) \\ B_i \setminus \varepsilon x(x \in (\mathbf{\Delta}_{i+1})) & \text{if } d(\alpha_{i+1}) \\ B_i \setminus \varepsilon x(x \in (\mathbf{\Delta}'_{i+1})) & \text{if } d'(\alpha_{i+1}) \\ B_i & \text{otherwise} \end{cases}$$

$$R_{i+1} = \begin{cases} R_i \cup \{\alpha_{i+1}\} & \text{if } r(\alpha_{i+1}) \\ R_i & \text{otherwise} \end{cases}$$

$$D_{i+1} = \begin{cases} D_i \cup \{\alpha_{i+1}\} \cup \varepsilon x(x \in (\mathbf{\Delta}_{i+1})) & \text{if } d(\alpha_{i+1}) \\ D_i \cup \{\alpha_{i+1}\} \cup \varepsilon x(x \in (\mathbf{\Delta}'_{i+1})) & \text{if } d'(\alpha_{i+1}) \\ D_i & \text{otherwise} \end{cases}$$

$$B_{\vdash}(\Theta, S) = B_k$$

$$B_{\vdash}(\Theta) = B_{\vdash}(\Theta, \varepsilon x(x \in \mathbf{S})).$$

The belief revision scheme mirrors the working of the TMS in an obvious way. This is why we speak of *TM belief revision* and *TM belief changes*. The definition of TM belief changes is less intricate than the algorithm of the TMS because it is defined on the assumption that the sequence of revision formulas conforms to a specific epistemic ranking. This does not mean that we cannot process new information that has lower epistemic rank than previously received information. But we have to recalculate the sets B_k, D_k, and R_k then, starting from B_i, D_i, and R_i, where α_{i+1} $(i < k)$ is the first element in the sequence S that has higher epistemic rank than the new revision formula. The problem of finding a rigorous definition of TM belief changes for sequences of revision formulas that fail to conform to a given epistemic ranking – and that avoids the detour of such a recalculation – must be left for future research.

We still need to explain some technical notions and symbols used in the TM belief revision scheme. $b(\alpha_{i+1})$, $d(\alpha_{i+1})$, and $r(\alpha_{i+1})$ are Boolean variables that, respectively, stand for α_{i+1} being believed, doubted and rejected. Note that $\mathbf{\Delta}_{i+1}$ is empty if $d(\alpha_{i+1})$ is false. Likewise, $\mathbf{\Delta}'_{i+1}$ is empty if $d(\alpha_{i+1})'$ is false.

A set $A \cup B$ is minimally inconsistent in B if and only if, for any $B' \subset B$, $A \cup B'$ is consistent, while $A \cup B$ is inconsistent. To put it in less formal terms, minimal inconsistency of $A \cup B$ in B means that every member of B is decisive for $A \cup B$ being inconsistent. Why should we doubt the beliefs in some $\Delta \subseteq (B_i)_{=\alpha_{i+1}}$ if both $d(\alpha_{i+1})$ is true and $(B_i)_{<\alpha_{i+1}} \cup \Delta \cup \{\alpha_{i+1}\}$ is minimally inconsistent in $\Delta \cup \{\alpha_{i+1}\}$? In this situation, we have an inconsistency between Δ and the new epistemic input α_{i+1}, in the context of $(B_i)_{<\alpha_{i+1}}$. The beliefs in $(B_i)_{<\alpha_{i+1}}$ have higher epistemic rank than α_{i+1}, while all members of Δ are on a par with $\{\alpha_{i+1}\}$. Since all members of $\Delta \cup \{\alpha_{i+1}\}$ are relevant for the inconsistency and since no such member is inferior to another, we must doubt all members of $\Delta \cup \{\alpha_{i+1}\}$. None of these members must be in the belief set of Θ.

In the definitions of $B_{\vdash}(\Theta)$, B_{i+1} and D_{i+1}, we make use of Hilbert's epsilon operator. This operator picks out an arbitrary object out of a set of objects that satisfy a certain property.

$$\varepsilon x \phi(x)$$

thus stands for some object x for which the proposition $\phi(x)$ is true. The epsilon operator was given a precise model-theoretic semantics in terms of choice functions.[10] A choice function simply picks out an object of a given set, for any subset of the domain of interpretation. On this semantics, all occurrences of $\varepsilon x(x \in A)$ designate the same object, relative to a given interpretation of the language with epsilon terms.

The semantics of choice functions give rise to three different notions of truth in the context of epsilon terms: first, we can simply accept that the evaluation of truth is relative to a choice function for sentences containing epsilon terms. The choice function may or may not be made explicit when asserting such sentences. Second, a sentence ϕ (possibly containing epsilon terms) is *generically* true if and only if ϕ is true for all choice functions. Third, a sentence ϕ (possibly containing epsilon terms) is *locally true* if and only if ϕ is true for at least one choice function.[11] For reasons that will become obvious shortly, we interpret our epsilon sentences in the sense of local truth. To be more precise, when devising an algorithm for the determination of the set $B_{\vdash}(\Theta)$, we understand the result of this algorithm according to the notion of local truth. For example, if an algorithm determines the set $B_{\vdash}(\Theta)$ such that there is a choice function such that $B_{\vdash}(\Theta) = B_{\vdash}(\Theta, \varepsilon x(x \in \mathbf{S}))$, then we say that the algorithm has correctly determined the set $B_{\vdash}(\Theta)$. If, by contrast, an algorithm determines the set $B_{\vdash}(\Theta)$ such that there is no choice function such that $B_{\vdash}(\Theta) = B_{\vdash}(\Theta, \varepsilon x(x \in \mathbf{S}))$, then we say that the algorithm has not correctly determined the set $B_{\vdash}(\Theta)$.

Why do we take recourse to epsilon terms and choice functions in the first place? Through the epsilon operator in the definition of D_{i+1}, we take into account that the

[10] This semantics has been developed by Asser [15]. For an accessible introduction to the epsilon calculus and its semantics, see Zach [121].

[11] See Meyer Viol [85] or Andreas and Schiemer [12] for the distinction between local and generic truth.

TMS may not detect all minimally inconsistent sets $(B_i)_{<\alpha_i} \cup \Delta \cup \{\alpha_{i+1}\}$ if $d(\alpha_{i+1})$ is true. As an example, take the following sequence of propositional formulas:

$$S = \langle p, p \to r, r \to q, u, u \to s, s \to r, \neg q \rangle.$$

Suppose the empty epistemic state is revised by these formulas in the order specified by sequence S. The TMS will then recognise the set $\{p, p \to r, r \to q, \neg q\}$ as a minimally inconsistent set, but not so the set $\{u, u \to s, s \to r, r \to q, \neg q\}$. For, the propositional constant r in the implication $r \to q$ is actually determined by the set $\{p, p \to r\}$, at the stage when the epistemic state is revised by $\neg q$. However, the propositional constant r in the implication $r \to q$ can also be determined by the set $\{u, u \to s, s \to r\}$, even in the resolution calculus. We cannot test for all possible determinations of a propositional constant in an application because this would lead to intractability. In the present example, this failure results in accepting the members of $\{u, u \to s, s \to r\}$ as beliefs. This is a problem since $\{u, u \to s, s \to r\}$ is not a subset of all maximal consistent subsets of $\{p, p \to r, r \to q, u, u \to s, s \to r, \neg q\}$. The underlying fundamental problem is that membership in all maximal consistent subsets is **coNP**-complete if the consistency test is tractable (cf. Sect. 7.1.2).

Likewise, if we were to interpret $B_{\vdash}(\Theta) = B_{\vdash}(\Theta, \varepsilon x(x \in \mathbf{S}))$ in the sense of generic truth, intractability would ensue. For, the TMS would have to determine the set $B_{\vdash}(\Theta, S)$ for all sequences $S \in \mathbf{S}$. When using Definition 8.13 for the determination of the set $B_{\vdash}(\Theta)$, we have to work with a specific sequence. Hence, we accept that the set $B_{\vdash}(\Theta)$ of actual beliefs may depend on the order of revisions. Any order that respects the epistemic ranking of Θ is fine.

Notice that the epsilon semantics of local truth does not result in understanding the set $B_{\vdash}(\Theta)$ such that $\phi \in B_{\vdash}(\Theta)$ iff there is a choice function such that $\phi \in B_{\vdash}(\Theta, \varepsilon x(x \in \mathbf{S}))$. For this to be seen, take the following example. Let A and B be two sets such that

$$A = \{\varepsilon x(x \in B)\}.$$

So, A is the singleton of the object picked out of set B by some choice function. If $B = \{a, b\}$, then both $A = \{a\}$ and $A = \{b\}$ are true, but not $A = \{a, b\}$. There is no choice function such that $A = \{a, b\}$.

The use of epsilon terms, thus, accounts for the fact that the revised set $B_{\vdash}(\Theta)$ of actual beliefs depends on the order of revision formulas. Alternatively, we could define $B_{\vdash}(\Theta, S) = B_k$, without eliminating S in the next step, thereby making the dependency on the order of revision formulas explicit. The benefit of using epsilon terms is simplicity in the subsequent definitions. In particular, we avail ourselves of an elegant way of saying that any sequence S that respects the epistemic ranking of Θ is admissible.

8.4.4 Tractability

We have defined the TM belief revision scheme in a general manner so that it is not bound to a specific logic. This revision scheme is, of course, not tractable for logics whose inference relation is not tractable. Let us therefore spell out TM belief changes for resolution frame logic. For this, we need to first explain the notion of a structuralist default theory in this logic:

Definition 8.14 ($\Theta_R(\mathbf{N})$) Let $\mathbf{N}$ be a structuralist theory-net. $\langle T_1, \ldots, T_n \rangle$ is a sequence of sets of $L_{RF}(\mathbf{N})$ sentences. Let $T_\cup = T_1 \cup \ldots \cup T_n$. $\Theta_R(\mathbf{N}) = \langle T_1, \ldots, T_n \rangle$ is a structuralist default theory in resolution frame logic if and only if

(1) for any $\gamma_\alpha \in T_\cup$, γ is a ground literal or clause of ground literals. If γ is a clause, $\bigvee \gamma$ is classically logically equivalent to an instance of the substantial law of $\mathbf{T}(\alpha)$.
(2) For all γ_α and all $\gamma'_{\alpha'}$, if $\mathbf{T}(\alpha) = \mathbf{T}(\alpha')$, then there is i $(1 \leq i \leq n)$ such that γ_α, $\gamma'_{\alpha'} \in T_i$, and $\gamma_\alpha, \gamma'_{\alpha'} \notin T_\cup \setminus T_i$.

Note that $\Theta_R(\mathbf{N})$ is a resolution-frame-logical variant of a structuralist default theory $\Theta(\mathbf{N})$, as defined by Definition 6.3. Condition (2) simply says that, if the applications of two $L_{RF}(\mathbf{N})$ sentences are applications of one and the same theory-element $\mathbf{T}$, then the two sentences must be at the same epistemic level of $\Theta_R(\mathbf{N})$. $\bigvee \gamma$ simply stands for some disjunction of the members of γ if γ is a clause. We are now in a position to define structuralist belief changes for resolution frame logic and the TM belief revision scheme:

Definition 8.15 ($K(\Theta_R(\mathbf{N}))$ **and** $K(\Theta_R(\mathbf{N})) *_i \alpha$) Let $\Theta_R(\mathbf{N})$ be a structuralist default theory of $\mathbf{N}$ in resolution frame logic. Let α be an $L_{RF}(\mathbf{N})$ sentence such that $\Theta_R(\mathbf{N}) \oplus_i \alpha$ is a structuralist default theory in resolution frame logic.

$$Inf(\Theta_R(\mathbf{N})) = Cn_{RF}(B_{\vdash_{RF}}(\Theta_R(\mathbf{N})))$$
$$K(\Theta_R(\mathbf{N})) = Inf(\Theta_F(\mathbf{N}))$$
$$K(\Theta_R(\mathbf{N})) *_i \alpha = Inf(\Theta_R(\mathbf{N}) \oplus_i \alpha)$$

where Cn_{RF} is the logical closure operation defined by $\models_{RF}$.

Let us further assume that every piece of disjunctive information forms a distinct application. To be more precise, for every application α, there is exactly one clause C_α in the set $T_1 \cup \ldots \cup T_n$ ($\Theta_R = \langle T_1, \ldots, T_n \rangle$) such that $|C_\alpha| > 1$, as suggested in Sect. 8.4.2. (Recall that this assumption is motivated by the problem of weak determinations.) Then, we can show tractability of TM belief changes for resolution frame logic. This can be done in two different ways. First, we can devise a resolution variant of the TMS in Chap. 7 that is sound and complete for resolution frame logic with respect to the TM belief revision scheme. Then, it follows from Theorem 7.1 that TM belief changes are tractable for resolution frame logic. This way of proving tractability is relevant for actual applications of the TMS, and will be worked out in the next section. In this section, however, we show tractability of TM belief changes in a more direct way, without relying on Theorem 7.1.

Theorem 8.8 *Let $\Theta_R(\mathbf{N})$ be a structuralist default theory in resolution frame logic. For every non-composite application α, there is exactly one clause C_α in $\Theta_R(\mathbf{N})$ such that $|C_\alpha| > 1$. Let ϕ be an $L_{RF}(\mathbf{N})$ sentence. Then, the problem of determining $B_{\vdash_{RF}}(\Theta_R(\mathbf{N}) \oplus_i \phi)$ has **P**-time complexity.*

This theorem says that TM belief changes are tractable for resolution frame logic if every piece of disjunctive information forms a distinct application. To prove the theorem, we need to show that there is an algorithm that determines $B_{\vdash_{RF}}(\Theta_R(\mathbf{N}) \oplus_i \phi)$ such that its computational complexity is polynomial. Let us adopt the following notational conventions:

- C_α: the clause such that C_α is an element of a member of $\Theta_R(\mathbf{N})$.
- $L_0(\alpha)$: the set of literals that represent presumed knowledge about the objects of α, prior to applying **T** and prior to exploiting links.
- $L(\alpha)$: the set of literals in the language $L(V(\alpha))$ that can be inferred from the union of $C_\alpha \cup L_0(\alpha)$ and the sets $L(\beta) \cup L_0(\beta)$ $(\beta \neq \alpha)$, but not from the union of $L_0(\alpha)$ and the sets $L(\beta) \cup L_0(\beta)$. When the algorithm starts, $L(\alpha)$ is set to the empty set for all applications in **N**.
- $C(\alpha)$: the logically strongest clause that can be inferred from the union of C_α, $L_0(\alpha)$ and the sets $L(\beta) \cup L_0(\beta)$ $(\beta \neq \alpha)$ (i.e., determinate literals of other applications) using the inference rules of resolution frame logic.

We shall see that $L(\alpha)$ is a singleton or empty, for any application α (Lemma 8.5). Let us now determine the computational complexity of an algorithm that determines $B_{\vdash_{RF}}(\Theta_R(\mathbf{N}) \oplus_i \alpha)$ according to Definition 8.13. Most of the recursive steps of this definition require testing the consistency of a set of $L_{RF}(\mathbf{N})$ sentences. We have already shown that this test is tractable on condition of a constraint on the syntactic size of applications (Theorem 8.3). We now show that the consistency test is tractable if every piece of disjunctive information forms a distinct application, even without a constraint on the syntactic size of applications.

Proposition 8.1 *Let Γ be a set of $L_{RF}(\mathbf{N})$ sentences. For any application α, there is exactly one clause $C_\alpha \in \Gamma$ such that $|C_\alpha| > 1$. Then, the test for satisfiability of Γ is **P**-time complex.*

This is an algorithm that tests for consistency of a set Γ of $L_{RF}(\mathbf{N})$ sentences on said condition.

Algorithm 8.2 (Satisfiability of a set Γ of $L_{RF}(\mathbf{N})$ sentences (assuming the premise of Proposition 8.1))

// Part A: determining intramodular inferences

$TR(\mathbf{N}) := \emptyset$; // $TR(\mathbf{N})$ is a dynamic set that collects those applications
// from which transmodular inferences may be drawn

for all $\alpha \in H(\mathbf{N})$:
  $C(\alpha) := \{l \mid l \in C_\alpha \text{ and } \bar{l} \notin L_0(\alpha)\}$
// $\bar{l}$ is the literal complementary to l.

```
    if |C(α)| = 1:
      L(α) := {l};
      TR(N) := TR(N) ∪ {α};
    else:
      L(α) = ∅;
      if C(α) = ∅:
        output: not satisfiable;
        return;

                // Part B: determining transmodular inferences

  repeat
    take some α ∈ TR(N):
      TR(N) := TR(N) \ {α};
      for all β ∈ H(N) \ {α}:
        if L̄(α) ⊆ C(β):                       // L̄(α) = {l̄} if L(α) = {l},
          C(β) := C(β) \ L̄(α);                 // L̄(α) = ∅ otherwise.
          if |C(β)| = 1:
            L(β) := C(β);
            TR(N) := TR(N) ∪ {β};
          if C(β) = ∅:
            output: not satisfiable;
            return;
  until TR(N) = ∅;
  output: satisfiable.
```

Part A of the algorithm implements the inference rule Intramodular Resolution, while Part B draws inferences according to the rule Transmodular Resolution. That the algorithm completely and correctly captures all intramodular inferences follows from Lemma 8.3 in the Appendix. Completeness and correctness of the determination of the transmodular inferences follow from the principle that only literals can be inferred from one application to another. If $|C(\alpha)| \neq 1$ and $|C(\beta)| \neq 1$, then there is no transmodular inference to be drawn from $\{C(\alpha), C(\beta)\}$. It is not difficult to show that the algorithm has **P**-time complexity. To be more precise:

Proposition 8.2 *Let all symbols and premises be as in Proposition* 8.1. *Let n be the number of applications in* **N**. *m is the maximal cardinality of a clause C_α such that C_α is a member of an element in $\Theta_R(\mathbf{N})$. Then, the test for consistency of the set Γ by Algorithm* 8.2 *has the computational complexity of $O(n^2 \cdot m)$.*

The proof is analogous to the demonstration that Algorithm 8.1 has **P**-time complexity, and can be found in the Appendix. Proposition 8.2 and the fact that the algorithm is complete and correct imply Proposition 8.1. Notably, if we assume that the maximal number of literals in a clause $C(\alpha)$ is bounded by a constant, we obtain $O(n^2)$ as computational complexity of the algorithm. This assumption is reasonable since we rarely, if at all, have generalisations whose instances require "very long" disjunctions in the present format.

To prove Theorem 8.8, we further need to show that – in case $d(\alpha_{i+1})$ is true – the problem of determining at least one member of the set $\mathbf{\Delta}_{i+1}$ is **P**-time complex. If $d(\alpha_{i+1})$ is false, $\mathbf{\Delta}_{i+1}$ is empty so that its determination is trivial. Likewise, for the set $\mathbf{\Delta}'_{i+1}$ relative to $d'(\alpha_{i+1})$. Without loss of generality, we show that the problem of determining at least one member of the set $\mathbf{\Delta}_{i+1}$ is **P**-time complex. Notice that we can keep track of the determinations among applications by introducing a binary relation $DT(\alpha, \beta)$ with the intended meaning that α determines β. Using the symbols of the above algorithm, we can say that α determines β if and only if $\overline{L(\alpha)} \subseteq C(\beta)$ and $|C(\beta)| > 1$. Further, we say that α determines β directly if and only if $DT(\alpha, \beta)$. α determines β indirectly if and only if $\neg DT(\alpha, \beta)$, but there are $\alpha, \alpha_1 \ldots, \alpha_j$ such that $DT(\alpha, \alpha_1)$ and for all i $(1 \le i < j)$ $DT(\alpha_i, \alpha_{i+1})$, and $DT(\alpha_j, \beta)$. With these concepts at hand, we can express and prove the following proposition:

Proposition 8.3 *Let Γ be a consistent set of $L_{RF}(\mathbf{N})$ sentences. For any application α, there is exactly one $C_\alpha \in \Gamma$ such that $|C_\alpha| > 1$. Let $DA(\beta)$ be the set of applications α such that a sentence l_α determines – directly or indirectly – a literal of β, as a consequence of running Algorithm* 8.2. *Let $\Gamma_{DA(\beta)}$ be the corresponding set of sentences: $\Gamma_{DA(\beta)} = \{\gamma_\alpha \mid \alpha \in DA(\beta)\}$. Suppose there is l_ε such that $\varepsilon \notin DA(\beta)$ and $\Gamma_{DA(\beta)}, l_\varepsilon \vdash_{RF} \Box$. But for any $\alpha \in DA(\beta)$ such that $\alpha \neq \beta$, we have $\Gamma_{DA(\alpha)}, l_\varepsilon \nvdash_{RF} \Box$. Also, $\Gamma_\beta, l_\varepsilon \nvdash_{RF} \Box$. Then, for any $\Gamma' \subset \Gamma_{DA(\beta)}$, it holds that $\Gamma', l_\varepsilon \nvdash_{RF} \Box$.*

This proposition makes an assertion as to which sentences γ_α are relevant for an inconsistency detected between two applications β and ε by Algorithm 8.2. Suppose an application β is found inconsistent with an application ε, where the inconsistency arises for the literals $\bar{l}_\beta$ and l_ε. Any set Γ_D of sentences such that D is a proper subset of the set of applications determining β is consistent with $\Gamma_{DA(\varepsilon)}$. Moreover, Γ_β is consistent with l_ε. Then, any member of the set $\Gamma_{DA(\beta)}$ is relevant for the inconsistency with l_ε.

Suppose Algorithm 8.2 is extended to keep track of the determinations among applications using a relation DT, as just indicated. Note that (i) keeping track of determinations can be done with only a constant number of additional computation steps. Further, note that if the extended algorithm detects an inconsistency between literals of two applications β and ε, we can read the sets $DA(\beta)$ and $DA(\varepsilon)$ off the relation DT. (ii) From these two sets we can easily determine the sets $\Gamma_{DA(\alpha)}$ and $\Gamma_{DA(\beta)}$. (This can be done using the general definition $\Gamma_A = \{\phi_\gamma \mid \phi_\gamma \in \Gamma$ and $\gamma \in A\}$.) Suppose the extended algorithm detects an inconsistency such that $C(\beta) = \{l\}$ and $C(\varepsilon) = \{\bar{l}\}$, where β and ε are believed, i.e., members of B_i. To apply Proposition 8.3 to the pair (β, ε), we need to also verify that any set Γ_D of sentences such that D is a proper subset of the set of applications determining β is consistent with $\Gamma_{DA(\beta)}$. (iii) This condition can be tested without additional computational costs. We merely need to ensure that, once a literal l_γ has been inferred by Intramodular Resolution, all transmodular inferences from this literal are drawn before any further inferences – involving literals other than l_γ – are drawn. Then, if some set Γ_D of sentences such that D is a proper subset of the set of applications determining β is inconsistent with Γ_ε, this inconsistency will be detected prior to detecting any inconsistency between β and ε. (We say that β is inconsistent with ε if and only if $C(\beta), C(\varepsilon) \vdash_R \Box$.)

(i), (ii), (iii) and Proposition 8.3 imply that at least one member of $\mathbf{\Delta}_{i+1}$ can be determined with only constant additional computation steps. From this and Proposition 8.2, we can infer:

Proposition 8.4 *Let all symbols and premises be as in Proposition* 8.1. *Let n be the number of applications in* **N**. *m is the maximal cardinality of a clause* C_α *such that* C_α *is a member of an element in* $\Theta_R(\mathbf{N})$. *The problem of determining the sets* B_{i+1}, D_{i+1}, *and* R_{i+1} *– on the basis of the sets* B_{i+1}, D_{i+1}, *and* R_{i+1} *– is in* $O(n^2 \cdot m)$.

We are now almost done. In order to determine $B_{\vdash_{RF}}(\Theta_R(\mathbf{N}) \oplus_i \alpha)$, we need to determine, first, the sets B_1 and R_1. This test consists in testing whether $\{\alpha_1\}$ is consistent. Then, we need to determine n times the sets B_{i+1}, D_{i+1}, and R_{i+1} on the basis of the sets B_i, D_i, and R_i. The set B_{n+1} directly gives us the set $B_{\vdash_{RF}}(\Theta_R(\mathbf{N}) *_i \alpha)$. From this and Proposition 8.4 it follows that:

Proposition 8.5 *Let all symbols and premises be as in Proposition* 8.1. *Let n be the number of applications in* **N**. *m is the maximal cardinality of a clause* C_α *such that* C_α *is a member of an element in* $\Theta_R(\mathbf{N})$. *The problem of determining the set* $B_{\vdash_{RF}}(\Theta_R(\mathbf{N}) \oplus_i \alpha)$ *is in* $O(n^3 \cdot m)$.

This proposition implies Theorem 8.8, and so we are done with the proof of this theorem.

It is important to note that the tractability result of Proposition 8.5 squares nicely with that of Theorem 7.1. The latter says that the computational complexity of the TMS algorithm is $O(n^2)$, where n is the number of applications in $H(\mathbf{N})$. Recall, moreover, that the problem of determining $B_{\vdash_{RF}}(\Theta_R(\mathbf{N}) \oplus_i \alpha)$ is in $O(n^3)$ if we assume that the maximal length of a disjunction is bounded by a constant. Such an assumption amounts to assuming an upper bound on the syntactic size of the modular units of reasoning, an assumption that has also been made in Theorem 7.1. On the assumption in question, the TMS algorithm has a lower computational complexity than an algorithm that directly implements Definition 8.13 using Algorithm 8.2 for the test of consistency. (We described such an algorithm when proving Proposition 8.5.) This should not come as a surprise. The algorithm underlying Proposition 8.5 and the TMS algorithm solve different, albeit related problems. The TMS exploits previously drawn inferences, given by the sets $H_s(\mathbf{N})$, $H_d(\mathbf{N})$, $H_r(\mathbf{N})$, $\mathbf{A}(\alpha)$, etc. It determines how these sets change when we receive a new epistemic input α. Proposition 8.5, by contrast, is obtained on the assumption that no previously drawn inferences are available so that the sets B_{n+1}, D_{n+1} and R_{n+1}, as well as the sets $C(\alpha)$, must be determined from scratch. Their determination starts with the determination of the sets B_1, D_1 and R_1. Theorem 7.1 and Proposition 8.5, therefore, tell us that truth maintenance pays off! The computational complexity drops from $O(n^3)$ to $O(n^2)$ when we keep track of previously drawn inferences and determinations. Not surprisingly, if we use the TMS for determining the set

$H_s(\mathbf{N})$ and the sets $\mathbf{A}(\alpha)$ from the scratch – starting from the empty belief base –, the corresponding algorithm has the computational complexity of $O(n^3)$.

Below, we shall describe and study a TMS algorithm that implements resolution frame logic. We will show that the resolution-based TMS has the computational complexity of $O(n^2)$ too (Theorem 8.9). This result even more directly establishes the claim that keeping track of previously drawn inferences and determinations allows for a more efficient computation of belief changes.

8.4.5 Comparison

Finally, let us indicate why the TM belief revision scheme does not suffer from the severe limitations of the extant tractable belief revision schemes surveyed in the Introduction (Sect. 1.8). Full meet revision, cut revision, linear revision, and some model-based belief revision schemes were found tractable for Horn logic there. Suppose the new epistemic input α is inconsistent with a set H of sentences, all of which are believed. Unlike the full meet revision scheme, the TM revision scheme does not force us to retract all beliefs β that are not logically true.

Unlike the cut revision scheme, there is no need to retract all beliefs that are not more firmly believed than the least firmly believed member of H. For example, when I realise that there is no butter in the fridge, I may well continue to believe that my bicycle is in the shed, even if the latter belief is not more firmly established than the previous belief that there is butter in the fridge. More generally, the TM belief revision scheme retracts a belief β only if β is relevant for an inconsistency with the new epistemic input.

Unlike linear revision, a TM belief revision does not assume a total epistemic order on the belief base. Unlike the tractable model-based belief revision schemes in Nebel [89, Sect. 4.1], TM belief revision works with an epistemic ordering on the belief base. In sum, the TM belief revision scheme is significantly more expressive and more realistic than the extant tractable belief revision schemes to be found in the literature.

8.5 Truth Maintenance by Resolution

Our first approach to thinking in frames was driven by model-theoretic semantic means, building upon the structuralist framework. The TMS devised in Chap. 7 determines and revises the admissible extensions of the applications of theory-elements. We shall now show how truth maintenance can be carried out in a proof-theoretic fashion. Instead of a natural deduction system of frame logic, however, we shall use resolution frame logic. Together with the TM belief revision scheme, this logic will enable us to devise a TMS that is sound, complete and tractable.

Why do we transform the semantic operations of the TMS into proof-theoretic ones? First, it is theoretically interesting to show that truth maintenance can be done by proof-theoretic means. Second, the problem of weak determinations does not arise in the first place when working with Resolution, as shown in Sect. 8.4.2. Third, for the proof-theoretic variant of the TMS, soundness and completeness can be shown in a relatively elegant way – relative to the semantics of resolution frame logic and the TM belief revision scheme. Fourth, some procedures of the TMS can be simplified using Resolution. To be more precise, some procedures of the proof-theoretic variant are computationally less demanding, while there is no procedure of this variant that is computationally more demanding than the original, semantic TMS. Since the original TMS has polynomial runtime, we can infer from this that the proof-theoretic TMS has polynomial runtime as well. Since the proof-theoretic TMS is sound and complete with respect to resolution frame logic and the TM belief revision scheme, this result confirms Theorem 8.8, according to which the TM belief revision scheme is tractable when used with resolution frame logic.

We do, of course, not claim that the resolution calculus gives us a cognitively more adequate account of human reasoning than natural deduction. However, it has been shown that a resolution-based proof can be translated into a natural deduction proof in polynomial time [45]. Hence, resolution-based operations of the proof-theoretic TMS correspond to natural deduction proofs, and the tractability of the former implies the tractability of the latter. So, we are using resolution for two reasons. First, to simplify the original TMS. Second, to ease the demonstration that there is a TMS that has all desired properties: polynomial runtime, soundness and completeness with respect to the semantics of frame logic and the TM belief revision scheme. Another benefit is that actual applications of the TMS in automated knowledge representation are easier to devise when working with Resolution.

8.5.1 Resolution-Based Truth Maintenance

Instead of working out all details of proof-theoretic truth maintenance, we shall describe the essential changes that are to be made in order to devise a variant of the TMS that is based on Resolution. Let us adopt the assumptions and notational conventions of the previous section. In particular, we assume that, for every non-composite application α, there is exactly one clause C_α in $\Theta_R(\mathbf{N})$ such that $|C_\alpha| > 1$. $L_0(\alpha)$: the set of literals that represent presumed knowledge about the objects of α, prior to applying $\mathbf{T}$ and prior to exploiting links. $C(\alpha)$: the logically strongest clause $l_1, \ldots, l_n$ that can be inferred from C_α, $L_0(\alpha)$ and determinate literals of other applications. $L(\alpha)$: the set of literals in the language $L(V(\alpha))$ that can be inferred from the union of C_α, $L_0(\alpha)$ and the sets $L(\beta)$, but not from the union of $L_0(\alpha)$ and the sets $L(\beta)$ alone.

Let us now describe the changes of our original TMS that are needed to obtain a resolution-based TMS: **revise_in**(α): instead of computing $\mathbf{A}(\alpha)$, determine $L(\alpha)$ and $C(\alpha)$ from $\{C_\alpha\} \cup \{L_0(\alpha)\}$ using Resolution. An algorithm for this determination is given by Part A of Algorithm 8.2. **revise**(α): this procedure is called only if $|C(\alpha)| = 1$ since only determinate literals can be inferred from one application to another. There is no need for a variable that has a function analogous to that of $\mathbf{A}_2(\alpha)$. The further changes are analogous.

satisfaction$(\alpha, \beta, cons)$: the core of **satisfaction**(α) in the resolution-based TMS is given by Part B of Algorithm 8.2. A given pair $(\alpha, \beta, cons)$ fails to pass the test for link satisfaction if and only if we happen to derive the empty clause from $\{C(\alpha), C(\beta)\}$. If so, $cons$ must be set to *false*.

The procedures **weakest**, **compare**, **reject** and **doubt** can be adopted without further changes. Also, the dynamic sets $H_s(\mathbf{N})$, $H_d(\mathbf{N})$, and $H_u(\mathbf{N})$ are still used to "collect" successful, doubted and unsuccessful applications, respectively. For example, if $C(\alpha), C(\beta) \vdash_R \Box$ – while $\alpha, \beta \in H_s(\mathbf{N})$ and α has a higher epistemic rank than β –, then β is moved from $H_s(\mathbf{N})$ to $H_u(\mathbf{N})$.

undetermine(Γ): instead of setting back $\mathbf{A}$ to $EXT(\mathbf{T}(\alpha))(\alpha) \cap \mathbf{M}(\mathbf{T})$, $L(\alpha)$ and $C(\alpha)$ are set back to their original values, which were assigned to these variables by **revise_in**(α). Thus, we have described a resolution-based variant of the original TMS. Henceforth, we refer to this variant by TMS_R.

No procedure of resolution-based truth maintenance is computationally more demanding than the corresponding procedure of the original TMS, which is based on model-theoretic means. Quite to the contrary: the costs of determining $L(\alpha)$ and $C(\alpha)$ are polynomial in the size of $|V(\alpha)|$, in the absence of composite applications. Intractability, therefore, will not arise from "large" applications α, large with respect to the vocabulary $V(\alpha)$ and the size of the domain of α. To achieve tractability for the original TMS, by contrast, we had to make the additional assumption that (i) $|EXT(T)(\alpha)|$ is bounded by a constant c_1, or (ii) $|EXT(T)(\alpha) \cap \mathbf{M}(\mathbf{T}(\alpha))|$ is bounded by a constant c_1 and the computational costs of determining $|EXT(T)(\alpha) \cap \mathbf{M}(\mathbf{T}(\alpha))|$ is bounded by a constant c_2. The tractability result of Theorem 7.1, therefore, carries over to the proof-theoretic variant – without any assumptions concerning the computational costs of determining $EXT(T)(\alpha) \cap \mathbf{M}(\mathbf{T}(\alpha))$. Thus, we obtain:

Theorem 8.9 *Let $\Theta_R(\mathbf{N})$ be a structuralist default theory in resolution frame logic. ϕ is an $L_{RF}(\mathbf{N})$ sentence. For every application α, there is exactly one clause C_α in $\Theta_R(\mathbf{N})$ such that $|C_\alpha| > 1$. Let n be the number of applications in $\mathbf{N}$. m is the maximal cardinality of a clause C_α such that C_α is a member of an element in $\Theta_R(\mathbf{N})$. Then, the computational complexity of the revision algorithm of TMS_R is $O(n^2)$.*

Proof. As just explained, this theorem follows from (i) Theorem 7.1 and (ii) no procedure of TMS_R is computationally more demanding than the corresponding procedure of the original TMS.

If truth maintenance by means of Resolution is simpler and computationally easier than by semantic means, why did we take the detour via the original, semantic TMS? A semantic approach to truth maintenance suggested itself from the strong semantic orientation of the structuralist framework. And we used this framework as starting point because it provided the formally most elaborated account of frame concepts to be found in the literature. More importantly, the semantics of resolution frame logic is motivated by the semantic approach to truth maintenance. Hence, we had the semantics of resolution frame logic in place when we set out to devise its proof theory. Likewise, for natural deduction frame logic devised in Sect. 8.2.

8.5.2 Compositions

So far, we have not taken into account composite applications, while assuming that all substantial laws of **N** are universally quantified implications. We shall not lift the latter assumption, but we can quite easily consider composite intended applications. If β_c is a composition of $\beta_1, \ldots, \beta_n$, then $\mathbf{C}(\beta_c)$ is given by the set $\{C_{\beta_1}, \ldots, C_{\beta_n}\}$, where C_{β_i} $(1 \leq i \leq n)$ is a clause such that $\bigvee C_{\beta_i}$ is logically equivalent to an instance of the substantial law of $\mathbf{T}(\beta_i)$. $L_0(\beta_c)$ is given by the set $L_0(\beta_1) \cup \ldots \cup L_0(\beta_n)$. $L(\beta_c)$ means what it means for non-composite applications, but this set may not be a singleton. The set $\mathbf{C}(\beta_c)$ is given by the set $\{C(\alpha_1), \ldots, C(\alpha_n)\}$. Recall that a composite application inherits all links from its components and remains linked to the component applications. Composite and component application must satisfy their obvious links. Note, however, that drawing inferences from $\mathbf{C}(\beta_c)$ is computationally more demanding than for non-composite applications. The computational costs of this are not polynomial in the size of $|V(\beta_c)|$. Composition, thus, leads to a gradual increase of the computational costs of truth maintenance. There are various algorithms that test for the satisfiability of a set of clauses, and so determine which literals are logical consequences of a set of clauses.[12] Using such a resolution-based algorithm, we can determine which literals of the language of a composite application can be inferred from $\mathbf{C}(\beta_c)$. These algorithms need exponential time in the worst case: the number of computation steps grows exponentially with the cardinality of the set of clauses that serves as premise set (see Cook and Reckhow [40] and Tseitin [111]).

Further, an important modification for composite applications is needed as regards the determination of frame-logical inferences within a composite application β_c. When $\mathbf{C}(\beta_c)$ has changed as a consequence of testing link satisfaction, we need to redetermine the intramodular inferences of $\mathbf{C}(\beta_c)$ using Intramodular Resolution, thereby possibly redetermining the set $L(\beta_c)$. For, the new information received from another application may allow us to draw further inferences within the application β_c. This is not necessary for applications α that are given be a single clause C_α of literals and a set $L_0(\alpha)$ of literals because of a lemma proven in the Appendix (Lemma 8.3). Furthermore, intramodular inferences of composite applications must not be drawn before drawing all intramodular and transmodular inferences of non-composite applications. Thereby, we ensure that any inconsistency detected between a composite application and another application is actually due to the formation of the composite application. On this condition, it

[12] See, e.g., Davis and Putnam [44], Davis et al. [43], Robinson [96] and Schöning [102, Chap. 2.5].

is justified to reject or doubt a composite application β_c, as well as its weakest components, if β_c is inconsistent with another application (composite or not) that has at least the epistemic rank of β_c. Finally, we should require compositions to be non-vacuous in the following sense:

Explanation 8.3 (Non-vacuous composition) Let β_c be an application that is composed of $\beta_1, \ldots, \beta_n$. Let $\mathbf{C}(\beta_c)$ be as just explained. β_c is non-vacuous if and only if there is exactly one literal γ such that $\mathbf{C}(\beta_c) \vdash_R \gamma$, while for any subset $\mathbf{C}' \subset \mathbf{C}(\beta_c)$, it holds that $\mathbf{C}' \nvdash_R \gamma$.

We impose this constraint on compositions in order to avoid problems analogous to the problem of irrelevant determinations, described in Sect. 7.4.5. Suppose β_c determines p to be true, in the sense that p cannot be inferred from the set of its components. Further, suppose β_c is found inconsistent with α that is successful and that has a higher epistemic rank than β_c. Then, β_c and its weakest components are rejected. If all components of β_c are inferentially relevant for the determination of p within β_c, then this operation is correct – with respect to the TM belief revision scheme. Otherwise, it may not be correct. For example, suppose one of the weakest components of β_c is inferentially non-relevant for the derivation of a literal l_{β_c}. Further, l_{β_c} is found inconsistent with some $\bar{l}_\alpha$, which can be derived from $C(\alpha)$ in the context of $L_0(\alpha)$ and sets $L(\varepsilon)$. Then, the inferentially non-relevant component of β_c is rejected. But this operation would not be correct with respect to the TM belief revision scheme. Hence, we require compositions to be non-vacuous.

8.5.3 Soundness, Completeness and Tractability

Frame logic has emerged out of an attempt at adjusting our logical theories of reasoning to the cognitive constraints of computational complexity. And the computational properties of the TMS in Chap. 7 have been used as an important heuristics to devise such a logical theory. Likewise, the TM belief revision scheme is motivated by an attempt at adjusting a specific belief revision scheme to the cognitive constraints of computational complexity. Furthermore, we have shown that carrying out truth maintenance by Resolution does not lead to an increase of computational complexity.

In light of these considerations, it should not come as a surprise that resolution-based truth maintenance exhibits – with respect to frame logic and the TM belief revision scheme – all the properties that we desire an algorithm for belief changes to have. That is, TMS_R provides us with a tractable computation of belief changes that is sound and complete with respect to frame logic and the TM belief revision scheme. To show soundness and completeness for TMS_R, let us be precise about the inferences that are actually drawn by this truth maintenance system:

Explanation 8.4 (Inference relation of TMS_R) Let γ be a literal and $\Theta_R(\mathbf{N})$ a structuralist default theory in resolution frame logic. γ is inferred from $\Theta_R(\mathbf{N})$ by

TMS_R if and only if there is an application α such that (i) $\gamma \in L(\alpha) \cup L_0(\alpha)$ and (ii) $\alpha \in H_S(\mathbf{N})$.

We are now in a position to assert and to show soundness and completeness for the resolution-based TMS:

Theorem 8.10 *Let $\Theta_R(\mathbf{N})$ be a structuralist default theory in resolution frame logic. For every non-composite application α, there is exactly one clause C_α in $\Theta_R(\mathbf{N})$ such that $|C_\alpha| > 1$. Further, any composite application is non-vacuous in the sense of Explanation* 8.3. *Then, TMS_R infers a literal γ from $(\Theta_R(\mathbf{N}) \oplus_i \phi)$ if and only if $B_{\vdash_{RF}}(\Theta_R(\mathbf{N}) \oplus_i \phi) \models_{RF} \gamma$.*

That is, TMS_R is sound and complete – with respect to resolution frame logic and the TM belief revision scheme – in the set of inferred literals. Most notably, this result opens up another way of establishing that the TM belief revision scheme is tractable for resolution frame logic – in addition to the proof of Theorem 8.8. By Theorem 8.9, we know that the computational complexity of the TMS_R algorithm is $O(n^2)$. Using this theorem, we can infer from soundness and completeness of TMS_R that the computational complexity of TM belief changes is $O(n^2)$ for resolution frame logic:

Theorem 8.11 *Let $\Theta_R(\mathbf{N})$ be a structuralist default theory in resolution frame logic. ϕ is an $L_{RF}(\mathbf{N})$ sentence. For every application α, there is exactly one clause C_α in $\Theta_R(\mathbf{N})$ such that $|C_\alpha| > 1$. Let n be the number of applications in* **N**. *There is a constant c such that, for any clause C_α that is a member of an element in $\Theta_R(\mathbf{N})$, $|C_\alpha| < c$. Then, the computational complexity of determining $B_{\vdash_{RF}}(\Theta_R(\mathbf{N}) \oplus_i \phi)$ is $O(n^2)$.*

This theorem simply follows from Theorems 8.9 and 8.10. Even if we allow compositions, TM belief changes remain tractable for resolution frame logic as long as the size of compositions is bounded by a constant:

Theorem 8.12 *Let $\Theta_R(\mathbf{N})$ be a structuralist default theory in resolution frame logic. ϕ is an $L_{RF}(\mathbf{N})$ sentence. For every non-composite application α, there is exactly one clause C_α in $\Theta_R(\mathbf{N})$ such that $|C_\alpha| > 1$. There is a constant c_1 such that, for any clause C_α that is a member of an element in $\Theta_R(\mathbf{N})$, $|C_\alpha| < c_1$. Any composite application is non-vacuous in the sense of Explanation* 8.3. *Further, the composition of applications in* **N** *is bounded by a constant c_2, in the sense that any composite application does not comprise more than c_2 applications. Then, the computational complexity of determining $B_{\vdash_{RF}}(\Theta_R(\mathbf{N}) \oplus_i \phi)$ is $O(n^2)$.*

Let us very briefly summarize the tractability results we obtained in this chapter. First, we have shown that finite natural deduction frame logic is fixed-parameter tractable, where the parameter is given by the syntactic size of the modular unit of reasoning (Theorem 8.3). Likewise, for resolution frame logic (Theorem 8.7). Then, we have shown that the TM belief revision scheme is tractable with respect to resolution frame logic when we assume a very fine-grained division of the given corpus of information into modular units of reasoning (Theorems 8.8 and 8.11). The crucial assumption is that every disjunction forms a distinctive unit of modular reasoning. On this assumption, the TM belief revision scheme is strictly tractable, i.e., without further assumptions about any parameter. We can overcome the limitation that every disjunction represents a distinctive unit of modular reasoning by forming compositions. The problem of determining TM belief changes in the presence of compositions is fixed-parameter tractable, where the crucial parameter is given by the number of components within a composite application (Theorem 8.12). In sum, frame logic and the TM belief revision scheme do in fact make up an account of dynamic tractable reasoning.

Chapter 9
Conclusions

Let us recall the specific problem that motivated our logical investigation of frames. On the one hand, there are cognitive processes that are inferential in nature: when we receive a new piece of information, we often draw inferences in the context of our old beliefs. We acquire new beliefs, while some old beliefs may be revised. On the other hand, computational complexity theory tells us that belief changes are computationally infeasible for human minds. This dilemma has been recognised as a major open problem in cognitive science by Cherniak [36], Fodor [58], Johnson-Laird [74], Stenning and van Lambalgen [107], Woods [120] and others. Carruthers [35] and Schneider [101] address the problem, but their solutions remain cursory. They invoke heuristics, which however remain largely unspecified. Likewise, there is no reasonably expressive and tractable belief revision scheme to be found in the computer science literature.[1]

Which of the premises that lead to the problem should be revised? First, one may deny that computational complexity theory is applicable to human minds, and thus deny that there is some substantial resemblance between computers and the biological machinery underlying our cognitive faculties. Computational complexity theory, however, is very liberal in its scope of application. Hence, we did not pursue this line of reasoning.

Second, it may be disputed that formal logic gives us a theoretical account of human reasoning. This seems too drastic a reaction. Such a move would not only break with a long-standing philosophical tradition, but also deprive us of any theoretical account of domain-general, non-probabilistic reasoning. Such an account is needed to explain how human minds are capable of reasoning about new domains of knowledge and inquiry. This is one of the reasons why Johnson-Laird [74] retains a broadly logical account of human reasoning from a cognitive science perspective.

[1]Cf. Nebel [89], who has given a comprehensive overview of complexity results in belief revision.

H. Andreas, *Dynamic Tractable Reasoning*, Synthese Library 420,
https://doi.org/10.1007/978-3-030-36233-1_9

Finally, we considered emending logical systems and belief revision schemes themselves, with the intent of devising formal systems of reasoning that are cognitively more adequate. This move suggests itself in light of the *cognitive turn* that modern logic is undergoing [112, p. 67]. By way of a cognitive emendation of logical systems, we continue to view the framework of formal logic as an (idealised) theory of human reasoning, and hold on to the computational paradigm of cognitive science.

Minsky [87] thought that logic-oriented approaches to human cognition are on the wrong track because they are not able to account for the cognitive units of memory, language and reasoning. Consequently, the notions of a frame and a frame concept have been advanced by Minsky as an alternative to a logical analysis of human cognition. We could show, however, that such frame concepts allow for a set-theoretic and logical representation: using set-theoretic predicates and some further elements of the structuralist representation scheme in the philosophy of science, we were able to characterise frame concepts in an axiomatic fashion.

As our topic is the cognitive feasibility of belief changes, we have then merged the axiomatisation of frame concepts with an AGM-style belief revision scheme. Thus, we obtained a structuralist account of belief changes. Based on this account, we devised a truth maintenance system, in the tradition of Doyle [47]. This truth maintenance system can be viewed as an algorithmic and semantic account of thinking in frames, i.e., reasoning with frame concepts.

The TMS (truth maintenance system) has been partially successful in establishing a tractable account of belief changes. One the one hand, it determines such changes in polynomial time, which means that tractability has been achieved. On the other hand, the TMS merely approximates the determination of belief changes – relative to finite, first-order logic and AGM-style approaches to belief revision.

In view of the tractability result of the TMS and the intractability of AGM-style belief changes, we took the approximate nature of the TMS to be a virtue rather than a failure. Accordingly, we derived some clues from the TMS about our means of coping with the computational and cognitive challenges of belief revision. A proper logic of modular reasoning thus emerged from an observation as to which types of inference are correctly and completely captured by the TMS. On this logic, full first-order reasoning is available within an application of a frame with a finite domain. For inferences from one frame application to another, however, only determinate literals are available. That is, only if a literal is determined to be true or false within an application of a frame concept, can it be "communicated" to another application.

We have given a natural deduction system of frame logic that is sound and complete with respect to a cognitively refined account of reasoning with frame concepts. This amounts to a proof-theoretic and syntactic account of reasoning and thinking in frames. Frame logic is reasonably expressive, but of course less expressive than first-order logic for finite domains. We have shown that inferential limitations of frame logic can be overcome by allowing for compositions of frame concepts. This leads to a gradual increase of the computational costs of determining consistency and logical consequence. Frame logic, however, remains tractable if we assume that the maximal size of a composition is bounded by a constant.

In addition to a natural deduction system, we devised a system of frame logic using the resolution calculus. This calculus allows for an even simpler implementation of the basic idea of frame logic: classical reasoning remains valid within a modular unit of reasoning, while only determinate literals may be inferred among such units. The inference rule Intramodular Resolution accounts for classical reasoning within a modular unit of reasoning. The inference rule Transmodular Resolution, by contrast, accounts for communication of determinate literals among modular units. This inference rule applies to pairs of formulas such that at least one element of the pair is a literal.

The main benefit of resolution frame logic – as compared to natural deduction frame logic – is simplicity. The resolution variant of frame logic is not only simpler in terms of the whole system of inference rules, but also in terms of the algorithmic implementation. This observation gave rise to a resolution-based TMS, which translates the semantic operations of the original TMS into syntactic operations. So, truth maintenance is first described in terms of semantic, model-theoretic operations, drawing on the structuralist framework. Then, it is described in a proof-theoretic fashion using the resolution calculus.

Finally, we have devised a belief revision scheme that is inspired by the truth maintenance algorithm. The key inspiration is to determine belief changes by way of a distinction between successful, doubted and unsuccessful applications of frame concepts. Thereby, certain inferences drawn at a certain stage of information can be exploited for the determination of further belief changes. Quite independently of computational constraints, this strategy is cognitively more plausible: when actually revising our beliefs, we do not start drawing inferences from a core of explicit beliefs from scratch, but evaluate the consistency of our explicit beliefs in light of previously drawn inferences. In particular, we do not directly determine maximal consistent subsets of sets of explicit beliefs. As the novel belief revision scheme is inspired by the working of the TMS, *TM belief revision* appeared a fitting label.

All these refinements eventually enable us to reach a state of harmony between our logical theories of belief change and the cognitive constraint of tractability. First, belief changes are tractable for resolution frame logic on the TM belief revision scheme. Second, the resolution-based TMS is sound and complete for resolution frame logic on this belief revision scheme. Thus, we have explained how dynamic reasoning is computationally feasible, thereby removing the major objection to the computational theory of mind. Moreover, we could advance a logical variant of the massive modularity hypothesis in cognitive science, which may help advance our understanding of human cognition in the foreseeable future.

Appendix A
Proofs

A.1 Chapter 5

Proposition 5.1 *For all sets* Γ *of* $L(V_o, V_t)$ *sentences and all sentences* ϕ *of* $L(V_o, V_t)$*:* $\Gamma \models_m \phi$ *if and only if* $\Gamma \models \phi$.

Forward: assume that $\Gamma \models_m \phi$, where the members of Γ and ϕ itself are $L(V_o, V_t)$ sentences. Proof by contradiction. Suppose $\Gamma \models \phi$ does not hold, i.e., ϕ is not a classical semantic consequence of Γ. Then, there is a structure $\mathcal{A}_1 = (D_{o1}, D_{t1}, \mathfrak{a}_{o1}, \mathfrak{a}_{t1})$ such that $\mathcal{A}_1 \models \Gamma$ but not $\mathcal{A}_1 \models \phi$. $\mathfrak{a}_{o1}$ and $\mathfrak{a}_{t1}$ interpret the observational and theoretical terms, respectively. Now, we need to construct a set $\mathbf{S}_{a1}$ of admissible structures such that $\mathcal{A}_1 \in \mathbf{S}_{a1}$. For this, take $\mathcal{A}_{o1} = (D_{o1}, \mathfrak{a}_{o1})$ as intended interpretation of $L(V_o)$. Then, let Γ_1 be the set of sentences that are members of Γ and have occurrences of theoretical terms. Further, take Γ_1 as the set of postulates for defining the set of admissible structures. Then, we obtain $\mathbf{S}_{a1} = MOD(\Gamma_1) \cap EXT(\mathcal{A}_{o1}, V_t, D_t)$. Since $\mathcal{A}_1 \in MOD(\Gamma_1)$ and $\mathcal{A}_1 \in EXT(\mathcal{A}_{o1}, V_t, D_t)$, it holds that $\mathcal{A}_1 \in \mathbf{S}_{a1}$. By assumption we know that $\Gamma \models_m \phi$. From this, $\mathcal{A}_1 \in \mathbf{S}_{a1}$ and $\mathbf{S}_{a1} \models_m \Gamma$, we can infer that $\mathcal{A}_1 \models \phi$. Contradiction.

Backward: assume that $\Gamma \models \phi$. Suppose, for contradiction, that $\Gamma \models_m \phi$ does not hold. So, there is a set $\mathbf{S}_{a1}$ of structures such that $\mathbf{S}_{a1} \models_m \Gamma$ but not $\mathbf{S}_{a1} \models_m \phi$. Hence, there is at least one structure $\mathcal{A}_1 \in \mathbf{S}_{a1}$ such that $\mathcal{A}_1 \models \Gamma$ but not $\mathcal{A}_1 \models \phi$. Therefore, $\Gamma \models \phi$ does not hold. Contradiction. □

Proposition 5.2 *Let* TC^R *designate the sentence that is obtained from* TC *by Ramsification, where* TC *is interpreted in the domain* $\langle D_o, D_t \rangle$*. Then,* TC^R *if and only if* $MOD(P_{TC}) \cap EXT(\mathcal{A}_o, V_t, D_t) \neq \emptyset$.

Forward: assume $(D_o, D_t, \mathfrak{a}_o) \models (TC)^R$, where $\mathcal{A}_o = (D_o, \mathfrak{a}_o)$ is the intended interpretation of $L(V_o)$. Then, there is an assignment of values to the second order variables of the open formula $TC(n_1, \ldots, n_k, X_1, \ldots, X_n)$ such that this formula is assigned to the value true. If this assignment is taken as an interpretation of the

H. Andreas, *Dynamic Tractable Reasoning*, Synthese Library 420,
https://doi.org/10.1007/978-3-030-36233-1

V_t terms using the correspondence between the variables of $(TC)^R$ and theoretical terms, while the V_o terms are interpreted according to $\mathcal{A}_o$, then every sentence of Γ_{TC} comes out true under this interpretation. Call the functions representing this interpretation of the V_t terms $\mathfrak{a}_{t1}$. Then, $(D_o, D_t, \mathfrak{a}_o, \mathfrak{a}_{t1})$ is an extension of the intended interpretation of $L(V_o)$ and also satisfying the postulates Γ_{TC}. Hence, $MOD(\Gamma_{TC}) \cap EXT(\mathcal{A}_o, V_t, D_t)$ is not empty.

Backward: assume that $MOD(\Gamma_{TC}) \cap EXT(\mathcal{A}_o, V_t, D_t)$ is not empty. Then, there is a structure $\mathcal{A} = (D_o, D_t, \mathfrak{a}_o, \mathfrak{a}_t)$ such that $\mathcal{A} \models \Gamma_{TC}$, where $\mathcal{A}_0 = (D_o, \mathfrak{a}_o)$ is the intended interpretation of $L(V_o)$. Then, obviously, there is an assignment to the second order variables in the open formula $TC(n_1, \ldots, n_k, X_1, \ldots, X_n)$ such that this formula is assigned to the value true, where the V_o symbols are interpreted according to $\mathcal{A}_o$. (Simply take the interpretation of the V_t terms, as given by the structure $\mathcal{A}$, as assignment to the second order variables using the correspondence between these variables and V_t terms.) Hence, $(D_o, D_t, \mathfrak{a}_o) \models (TC)^R$. □

A.2 Chapter 7

Proposition 7.1 *If $K(\Theta)$ is closed under $\mid\!\sim^{PS(s)}$, the problem of deciding if $\phi \in K(\Theta) \oplus_i \alpha$ is* **coNP^NP**-*complete in propositional logic.*

For the sake of conciseness and simplicity, we show this proposition using theorems in Nebel [89]. A self-contained exposition of the complete proof would require a much more involved exposition of both belief base revision theory and computational complexity. Theorem 11 in Nebel [89] says that prioritised belief base revision is $\mathbf{coNP}^{\mathbf{NP}}$-complete for propositional logic. Prioritised base revision is a close relative to the belief revision scheme with preferred subtheories. There is no difference, at the level of formal properties, between a prioritised belief base $\mathcal{H}$ and a default theory Θ. Both are sets of formals furnished with a weak ordering.

Let us first assume that the revision formula is not a logical falsehood. The proof of the hardness part of Theorem 11 in Nebel [89] is based on Theorem 9 in Nebel [89], which says that full meet base revision is $\mathbf{coNP}^{\mathbf{NP}}$-complete. More precisely, it is based on the claim that full meet base revision is a special case of prioritised base revision, viz., the case where we have only one epistemic level. Let H be a belief base. Full meet base revision with α is equivalent to determining all maximal consistent subsets S of H such that $S \not\models \neg\alpha$, and test if $S, \alpha \models \phi$ for all such subsets. Note that $S \not\models \neg\alpha$ iff $S \cup \{\alpha\}$ is satisfiable. Hence, we can use the PSA formalism to determine these subsets S of H by determining the preferred subtheories of $\langle H \rangle \oplus_0 \alpha$. ($\langle H \rangle \oplus_0 \alpha = \langle \{\alpha\}, H \rangle$.) Testing if $S, \alpha \models \phi$ is straightforward if we can use an oracle for satisfiability in propositional logic. Hence, full meet base revision is reducible to belief revision with preferred subtheories (using sceptical reasoning) with polynomial overhead. Theorem 9 in Nebel [89] therefore implies that deciding if $\phi \in K(\Theta) \oplus_i \alpha$ is $\mathbf{coNP}^{\mathbf{NP}}$-hard.

Let us move on to showing that deciding if $\phi \in K(\Theta) \oplus_i \alpha$ is a member of $\mathbf{coNP}^{\mathbf{NP}}$. The algorithm used by Nebel [89] to show that prioritised belief base revision is in $\mathbf{coNP}^{\mathbf{NP}}$ can be modified to solve the problem of deciding if $\phi \in K(\Theta) \oplus_i$ without changing the complexity of this algorithm. Instead of testing if $B \in A \downarrow \neg\alpha$, we insert α at level i into the prioritised belief base and test if $B \in A \downarrow \bot$.

There remains the case where the revision formula α is a logical falsehood. The problem of prioritised belief base revision is trivial in this case since it holds, for any formula ϕ, $\phi \in \mathcal{H} * \alpha$. The problem of deciding if $\phi \in K(\Theta) \oplus_i \alpha$ remains non-trivial, but the fact that α is a logical falsehood does not change the computational complexity of the algorithm. For, even if the revision formula is a logical falsehood, we need to construct maximal consistent subsets of the default theory Θ. □

Proposition 7.2 *If $K(\Theta)$ is closed under $\mid\!\sim^{PS(su)}$, the problem of deciding if $\phi \in K(\Theta) \oplus_i \alpha$ is $\mathbf{coNP}^{\mathbf{NP}}$-complete in propositional logic.*

Theorem 26 in Nebel [89] says that, for prioritised belief base revision, computational complexity does not change if we take the intersection of the set of inclusion-maximal subsets of the prioritised belief base $\mathcal{H}$ and test for this intersection entailment of ϕ instead of testing if ϕ is entailed by all inclusion-maximal subsets of the prioritised belief base $\mathcal{H}$. In the proof of Proposition 7.1, we have used reductions between algorithms that, respectively, underlie prioritised belief base revision and belief revision with PSA. Using these reductions, we can infer Proposition 7.2 from Theorem 26 in Nebel [89]. □

Proposition 7.3 *Any branch of an operation tree of the TMS has the following property. If the operation* **revise**(α_{i+1}) *is the next revision operation that follows the operation* **revise**(α_i) *– in the sense of an operation sequence of type* (7.1) *–, then the epistemic rank of α_{i+1} is at most as high as that of α_i.*

As for the sequence of operations between **revise**(α_i) and **revise**(α_{i+1}), the following cases are to be distinguished:

(1) $\langle rev(\alpha_i), det(\alpha_{i+1}), rev(\alpha_{i+1})\rangle$
(2) $\langle rev(\alpha_i), doubt(\beta), rev(\alpha_{i+1})\rangle$
(3) $\langle rev(\alpha_i), doubt(\beta), undet(\gamma_1), \ldots, undet(\gamma_k), rev(\alpha_{i+1})\rangle, k \geq 1$
(4) $\langle rev(\alpha_i), rej(\beta), rev(\alpha_{i+1})\rangle$
(5) $\langle rev(\alpha_i), rej(\beta), undet(\gamma_1), \ldots, undet(\gamma_k), rev(\alpha_{i+1})\rangle, k \geq 1$.

Case (1): Since α_{i+1} is determined by α_i, it cannot have a higher rank than α_i. For, the epistemic rank of α is given by the rank of the theory-element of the epistemically weakest intended application that determines α. More precisely, the epistemic rank of α is given by the rank of the theory-element $\mathbf{T}(\psi)$, where ψ is some member of $DI_w(\alpha)$.

Case (2): If some β is doubted, all intended applications γ that are rejected because of inconsistencies with β must be reconsidered. That is, **revise** must be called for all these γ. Now, (i) the epistemic rank of β is equal or lower than that of α_i if $\beta \in H_s(\mathbf{N})$ is doubted by $\alpha_i \in H_s(\mathbf{N}) \cup H_d(\mathbf{N})$. (ii) All $\gamma \in B(\beta)$ are inferior to β since this is a precondition of having been rejected by β. Because of (i) and (ii), all γ (for which an operation **revise** is called) are epistemically inferior to α_i. Hence, for $\gamma = \alpha_{i+1}$ it holds that the epistemic rank is lower than that of α_i.

Case (3): If some β is doubted in the course of an operation **revise**(α_i), determinations by β must be retracted. Let γ be an intended application whose determination by β is retracted. After the retraction, **revise**(δ) will be called for all δ that are blocked by γ. γ blocking δ, however, presupposes that (i) $\gamma \leq \delta$ prior to the undetermination of γ. Further, (ii) $\beta \leq \gamma$ if β determines γ. Since α_i doubts β, (iii) $\alpha_i \leq \beta$. (i), (ii) and (iii) imply that $\alpha_i \leq \delta$. $\delta = \alpha_{i+1}$ as **revise**(δ) is the next call of **revise** that follows **revise**(α_i). Hence, α_{i+1} cannot be epistemically superior to α_i in our sequence of operations.

Case (4): This case is analogous to case (2), except for the fact that β is inferior to α_i. All intended applications that had been blocked by β and for which thus **revise** must be called are therefore inferior to α_i.

Case (5): This case is analogous to case (3), except for the fact that β is inferior to α_i. □

Proposition 7.4 *There is no branch that has one of the following types as sub-branch:*

$$\langle \ldots, doubt(\alpha), \ldots, rev(\alpha), \ldots \rangle$$

$$\langle \ldots, rej(\alpha), \ldots, rev(\alpha), \ldots \rangle.$$

Let us begin with ruling out the first type of sequence. Suppose α has been doubted by some β. Then, **revise**(α) will only be called if β becomes rejected by some γ that is epistemically superior to β. This, however, presupposes a call of **revise** for γ within the branch in question. Hence, there would have to be a sequence

$$\langle rev(\alpha), \ldots, rev(\gamma), \ldots \rangle$$

such that γ has a higher epistemic rank than α, or a sequence

$$\langle rev(\beta), \ldots, rev(\gamma), \ldots \rangle$$

such that γ has a higher epistemic rank than β. This, however, is impossible according to Proposition 7.3.

The second type of sequence can be ruled out in a similar vein. The only difference is that we have an application β rejecting α such that β must have a higher epistemic rank than α – instead of an application β that doubts α, where α and β have the same epistemic rank. Hence, a second call of **revise** for α presupposes that some γ rejects or doubts β, which implies that γ has at least the epistemic rank of β. Using $\alpha < \beta$, we can infer from this that $\alpha < \gamma$. This, however, is impossible according to Proposition 7.3. □

Proposition 7.5 *Any branch of an operation tree is finite.*

By Proposition 7.4, we know that there are no two calls of **revise** for one and the same application α within one branch. Hence, if we have a call of **revise** for α_{i+1} that is the next call of **revise** following upon **revise**(α_i), it must hold that $\alpha_{i+1} \neq \alpha_i$. Therefore, in order to show that any branch of the operation tree is finite, we need to show that the number of operations between two calls of **revise** is always finite. As for the sequence of operations between **revise**(α_i) and **revise**(α_{i+1}), the following cases are to be distinguished:

(1) $\langle rev(\alpha_i), det(\alpha_{i+1}), rev(\alpha_{i+1})\rangle$
(2) $\langle rev(\alpha_i), doubt(\beta), rev(\alpha_{i+1})\rangle$
(3) $\langle rev(\alpha_i), doubt(\beta), undet(\gamma_1), \ldots, undet(\gamma_k), rev(\alpha_{i+1})\rangle, k \geq 1$
(4) $\langle rev(\alpha_i), rej(\beta), rev(\alpha_{i+1})\rangle$
(5) $\langle rev(\alpha_i), rej(\beta), undet(\gamma_1), \ldots, undet(\gamma_k), rev(\alpha_{i+1})\rangle, k \geq 1$.

Case (1): Obviously, the number of operations between two successive calls of **revise** is finite in this case.

Case (2): As in case (1), the number of operations between two successive calls of **revise** is finite.

Case (3): If an application β becomes doubted by α_i, then the determinations of β need to retracted. This may lead to retracting further determinations. Could there be an infinite chain of undeterminations? This is not a possibility for the following reasons. Retracting some determination of α means extending the dynamic set $\mathbf{A}(\alpha)$. Since, however, $|EXT(\mathbf{T})(\alpha) \cap \mathbf{M}(\mathbf{T})|$ is finite for any application $\alpha \in H(\mathbf{N})$, undeterminations must come to an end after a finite number of steps. Hence, there cannot be infinite chains of undetermination.

Case (4): This case is trivial as the cases (1) and (2) are.

Case (5): This case is analogous to case (3). □

Proposition 7.6 *For any node of an operation tree, the number of immediate subnodes is smaller than or equal to* $2|H(\mathbf{N})|$.

Five cases need to be distinguished. The number of operations – which are represented by edges and corresponding subnodes – following upon an operation

(1) $revise(\alpha)$
(2) $det(\beta)$
(3) $doubt(\beta)$
(4) $reject(\beta)$
(5) $undet(\beta)$.

Case (1). Executing an operation *revise*(α) consists in testing link satisfaction for all pairs (α, β), $\beta \neq \alpha$ and $\beta \in H_s(\mathbf{N}) \cup H_d(\mathbf{N})$. Each such test can induce the following types of operation: (i) a determination on α or β or both, (ii) a doubt on α or β or both, (iii) a rejection of α or β, (iv) no such operation at all. For obvious reasons, α cannot be doubted or rejected twice as an immediate consequence of **revise**(α). α, however, may well become determined more than once. Hence, each test of link satisfaction with some $\beta \in H_s(\mathbf{N}) \cup H_d(\mathbf{N})$ within **revise**(α) induces at most two operations, which amounts to two edges with corresponding subnodes. (i) The maximal number of subnodes of a node **revise**(α), therefore, is smaller than or equal to $2(|H(\mathbf{N})| - 1)$.

Case (2): Any determination is followed by exactly one revision.

Case (3): The maximal number of applications revised or undetermined (as a consequence of an operation of doubt of β) is smaller than $|H(\mathbf{N})|$. For, it is only applications that are blocked or directly determined by β that need to be revised or undetermined, respectively, after β has been doubted.

Case (4): This case is analogous to case (3).

Case (5): Likewise, the operation of undetermining γ requires retraction of the direct determinations of γ and reconsideration of those applications blocked by γ. The number of applications directly determined by γ or blocked is certainly smaller than $|H(\mathbf{N})|$. Hence, the number of immediate subnodes of a node that represents undetermination of an application is smaller than $|H(\mathbf{N})|$. Hence, in all five cases, the number of immediate subnodes is smaller than or equal to $2|H(\mathbf{N})|$. □

Proposition 7.8 *If an operation* ***revise***(α_{i+1}) *is executed after an operation* ***revise***(α), *the epistemic rank of* α_{i+1} *cannot be higher than that of* α.

We know that the determination of revisions begins with an operation **revise**(α). Recall, moreover, that further calls of **revise** (that follow upon the initial call **revise**(α)) are prioritised in such a manner that those applications with the highest epistemic rank are taken into account first. In other words, the TMS first collects the intended applications to be revised until all operations of determination, doubt, undetermination and reject are executed that are direct consequences of **revise**(α) – in the sense that these operations are not consequences of an operation **revise**(β), $\alpha \neq \beta$. Second, the applications thus collected are executed according to their epistemic rank, where the order of execution does not matter for applications of equal epistemic rank.

We prove our proposition by induction. Induction base: suppose **revise** has been called for the initial revision application α. As a consequence of the initial revision operation **revise**(α), we obtain a totally ordered set $TR = \{\alpha_1, \ldots, \alpha_k\}$ of applications for which **revise** needs to be called. By Proposition 7.3, we know that if a branch contains the sequence

$$\langle rev(\alpha), op_{1,1}, \ldots, op_{1,k_1}, rev(\alpha_i), \ldots \rangle$$

the epistemic rank of α_i cannot be higher than that of α. Hence, none of the applications $\alpha_i \in TR$ has a higher epistemic rank than the initial revision application α. Therefore, it holds for the revision operation **revise**(α_i) executed next to **revise**(α) – in the sense that no other revision operation between **revise**(α) and **revise**(α_i) is executed – that α_i has not a higher epistemic rank than α.

Induction step: suppose we have a totally ordered set $TR = \{\alpha_1, \ldots, \alpha_n\}$ of applications for which **revise** needs to be called. Let α_i be among the applications of the highest epistemic rank and let **revise**(α_i) be the revision operation called next. As a consequence of this call, the dynamic set TR is expanded by a set $\{\alpha_{n+1}, \ldots, \alpha_{n_1}\}$. By Proposition 7.3 we know that none of these applications has a higher epistemic rank than α_i. Hence, it holds for the revision operation **revise**(α_j) executed next to **revise**(α_i) that α_j has not a higher epistemic rank than α_i. This concludes the inductive proof. □

Proposition 7.9 *For all intended applications α, we can have at most one operation **doubt**(α) or **reject**(α), but not both operations, in one and the same operation tree.*

We prove this proposition by contradiction. Case (1): suppose (i) α has been doubted by β_1. Assume, further, that (ii) α will be doubted a second time. The operation of doubting α by β_1 must be an immediate consequence of an operation **revise**(β_1) or **revise**(α). Now, (ii) presupposes that (iii) some member of $DI_p(\beta_1)$ becomes rejected by an application γ after α has been doubted by β_1. For, α can only be doubted a second time if it becomes undoubted after the first operation **doubt**. Undoubting of α, however, requires rejection of some member of $DI_p(\beta_1)$, the set of applications that determine β_1. (iii) implies that (iv) an operation **revise**(γ) is executed later than an operation **revise**(α) or an operation **revise**(β_1), where γ must have a higher epistemic rank than both β_1 and α (which have equal epistemic rank). If γ did not have a higher epistemic rank, it could not reject some member of $D_p(\beta_1)$. By Proposition 7.3, we know that (iv) is not possible; contradiction. Likewise, once doubted by an application β_1, α cannot be rejected any more in one and the same operation tree. For this presupposes that (v) link satisfaction is tested for a pair (α, γ) such that $\gamma \in H_s(\mathbf{N})$ and γ has a higher epistemic rank than α. If α has been doubted by β_1 as an immediate consequence of an operation **revise**(α), (v) is not possible because of Proposition 7.3. If α has been doubted as an immediate consequence of an operation **revise**(β_1), (v) is not possible because this would imply that γ has a higher epistemic rank than β_1, which is impossible according to Proposition 7.3.

Case (2): suppose (i) α has been rejected by β_1. In a manner analogous to case (1), we can show that α can then neither be doubted nor rejected after that operation in one and the same operation tree. □

Proposition 7.10 *For all intended applications α, we can have at most $(|H(\mathbf{N})|-1)$ operations **undetermine**(α), in an operation tree.*

Obviously, an operation **undetermine**(α) presupposes a preceding operation **doubt**(β) or **reject**(β). Also, an operation **doubt**(β) or **reject**(β) cannot immediately initiate more than one operation **undetermine**(α). From Proposition 7.9 we can infer that we can have at most $|H(\mathbf{N}) - 1|$ operations of the type **reject** or **doubt** in an operation tree. These observations together imply that we can have at most $(|H(\mathbf{N})| - 1)$ operations **undetermine**(α), in an operation tree. □

Proposition 7.11 *For all intended applications α, we can have at most $(CM_m - 1)(|H(\mathbf{N})| - 1)$ determinations of α, in an operation tree.*

Suppose the applications determining α do not become rejected or doubted at a later stage. There are at most $|EXT(\mathbf{T})(\alpha) \cap \mathbf{M}(\mathbf{T})| - 1$ determinations of

α, then. For, determining an application consists in diminishing the (dynamic) set $\mathbf{A}(\alpha)$. This process comes to an end if $\mathbf{A}(\alpha) = 1$. Determinations, however, may become retracted at a later stage, which allows for a redetermination. Since, however, retraction requires doubting or rejecting a determining application and since there are at most $|H(\mathbf{N})| - 1$ operations of **doubt** or **reject**, there cannot be more than $(|EXT(\mathbf{T})(\alpha) \cap \mathbf{M}(\mathbf{T})| - 1)$ $(|H(\mathbf{N})| - 1)$ operations of determinations of an application α. Therefore, there cannot be more than $(CM_m - 1)(|H(\mathbf{N})| - 1)$ operations of determination of an application α. □

Proposition 7.12 *For all intended applications α, the number of operations **revise**(α) in an operation tree is smaller than or equal to* $(1 + CM_m)|H(\mathbf{N})|$.

There are four occasions for a call of **revise** for α. Case (1): the initial revision operation. Case (2): as a consequence of an operation **determine**(α). Case (3): as a consequence of undetermining an application β that has doubted or rejected α. Case (4): as a consequence of rejecting an intended application β that has rejected or doubted α.

Obviously, there can be at most one call of **revise** of type (1).

Let us move on to calls of **revise** of type (2). By Proposition 7.11, we cannot have more than $(CM_m - 1)(|H(\mathbf{N})| - 1)$ determinations of α. Hence, we can have not more than $(CM_m - 1)(|H(\mathbf{N})| - 1)$ calls of **revise** for α of type (2).

Type (3). By Proposition 7.10, we can have at most $(|H(\mathbf{N})| - 1)$ operations of the type **undetermine**. Hence, we can have at most $(|H(\mathbf{N})| - 1)$ operations **revise**(α) of type (3).

Type (4). By Proposition 7.9, any application can doubt or reject an application α only once. Hence, any application can be undoubted or unrejected only once. Hence, we can have at most $|H(\mathbf{N})|$ operations **revise**(α) of type (4).

In sum, we cannot have more than $(1 + CM_m)|H(\mathbf{N})|$ calls of **revise** for an intended application α in an operation tree. □

A.3 Chapter 8

Theorem 8.1 (Soundness) *Let Γ be a set of $L_F(\mathbf{N})$ sentences and let ϕ be such a sentence. If $\Gamma \vdash_F \phi$, then $\Gamma \models_F \phi$.*

Proof. Soundness can be proven in the standard way, i.e., by using an inductive proof on the number of inferences of a derivation. For the induction base, suppose the number of inferences is zero. Then, the derivation has the form $\{\phi\} \vdash_F \phi$. The tree that consists of just a single sentence trivially satisfies the definition of a derivation (Definition 8.3). Furthermore, $\{\phi\} \models_F \phi$ holds trivially if ϕ is an $L_F(\mathbf{N})$ sentence.

Induction step. Let Γ' be the set Γ united with all singletons $\{\phi\}$ such that ϕ is a premise of a so far uncompleted subproof. Let n be the number of inferences drawn so far from sentences of Γ'. By the induction hypothesis, we know that (i) all so far derived sentences are logical consequences – in the sense of $\models_F$ – of Γ'. Now, we need to show that (ii) any correct application of an inference rule leads to a sentence γ such that γ is a frame-logical consequence of the set of so far derived sentences, where the derivation is at step n. (i) and (ii) imply that γ is a frame-logical consequence of Γ'. So, we need to establish proposition (ii) for each inference rule of resolution frame logic.

Let us begin with the inference rules of intramodular reasoning, i.e., the inference rules that result from corresponding rules in standard systems of natural deduction through relativisation to an application or a set of applications. For these inference rules, proposition (ii) can be proven in a manner that is closely analogous to corresponding proofs for the natural deduction rules of classical first-order logic.[1] Recall that, within a modular unit of reasoning, first-order reasoning remains valid. This is why the proofs of proposition (ii) for the natural deduction rules of classical first-order logic can be adopted with only minor and straightforward modifications.

For the inference rules that govern the scope of assertions – Scope Restriction, Instantiation, Comprehension and Extension –, proposition (ii) follows from the semantics of $L_F(\mathbf{N})$ sentences, which is given by the Definitions 8.4–8.6. Likewise, for Law Application 1 and 2 (which allow us to assert the substantial law of $\mathbf{T}$ for all applications of $\mathbf{T}$), proposition (ii) follows from the semantics of $L_F(\mathbf{N})$ sentences, according to which the substantial law of $\mathbf{T}$ must be verified by all admissible extensions of an application α of $\mathbf{T}$, on all admissible interpretations of $\mathbf{AE(T)}$.

Finally, we have the two inference rules of transmodular reasoning, Link Exploitation 1 and 2. Suppose $R(c_1, \ldots, c_n)$ is verified by all admissible extensions of $\mathbf{I}(\alpha)$ on all admissible interpretations of $\mathbf{AE(T)}$. Then, by the postulate schema $\mathbf{P}4.1(\mathbf{T})$, $R(c_1, \ldots, c_n)$ must be verified by all admissible extensions of $\mathbf{I}(\beta)$ on all admissible interpretations of $\mathbf{AE(T)}$ if $R(c_1, \ldots, c_n) \in L(V(\beta))$. Hence, proposition (ii) holds for Link Exploitation 1. The demonstration of proposition (ii) for Link Exploitation 2 works analogously. □

Theorem 8.2 (Completeness) *Let Γ be a set of $L_F(\mathbf{N})$ sentences. ϕ is a quantifier-free sentence of $L_F(\mathbf{N})$. Suppose $\vdash_F$ and $\models_F$ form a system of finite frame logic in the sense of Explanation 8.2. If $\Gamma \models_F \phi$, then $\Gamma \vdash_F \phi$.*

Suppose $\Gamma \models_F \phi$. This means that ϕ is verified by all admissible interpretations of $\mathbf{AE(T)}$ (for all $\mathbf{T} \in \mathbf{N}$) that verify all members of Γ. Let us distinguish between a set $P_1(\mathbf{N})$ that contains the postulates $P_F(\mathbf{T}) \setminus \{P4.1(\mathbf{T}), P4.2(\mathbf{T})\}$ for all $\mathbf{T} \in \mathbf{N}$ and a set $P_2(\mathbf{N})$ that contains $P4.1(\mathbf{T})$ and $P4.2(\mathbf{T})$ for all $\mathbf{T} \in \mathbf{N}$. $P_1(\mathbf{N})$ defines the semantics of intramodular reasoning, whereas $P_2(\mathbf{N})$ that of transmodular reasoning. Thus, we can distinguish between two types of set of admissible $L(\mathbf{N})$ structures (admissible in the sense of Definition 5.7):

[1] These proofs can be found in various advanced textbooks of mathematical logic (see, e.g., Zach [122]).

$$\mathbf{S}_{a,i} = MOD(\Gamma) \cap MOD(P_1(\mathbf{N})) \cap MOD(D(\mathbf{N})) \cap EXT(\mathcal{A}_o, V_t(\mathbf{N}) \cup V_d(\mathbf{N}))$$

$$\mathbf{S}_{a,t} = MOD(\Gamma) \cap MOD(P_2(\mathbf{N})) \cap MOD(D(\mathbf{N})) \cap EXT(\mathcal{A}_o, V_t(\mathbf{N}) \cup V_d(\mathbf{N})).$$

These two types of sets of admissible structures give rise to two different relations of logical consequence:

$$\Delta \models_{F,i} \psi \text{ iff, for all } \mathcal{A} \in \mathbf{S}_{a,i} \text{ if } \mathcal{A} \models_F \alpha \text{ for all } \alpha \in \Delta, \text{ then } \mathcal{A} \models_F \psi.$$

$$\Delta \models_{F,t} \psi \text{ iff, for all } \mathcal{A} \in \mathbf{S}_{a,t} \text{ if } \mathcal{A} \models_F \alpha \text{ for all } \alpha \in \Delta, \text{ then } \mathcal{A} \models_F \psi.$$

Using the semantic constraints of finite frame logic, we show that there are sets Δ and Δ' such that

$$\Gamma \models_{F,i} \Delta$$

$$\Delta \models_{F,t} \Delta'$$

$$\Delta' \models_{F,i} \phi$$

where $\Gamma \models_{F,i} \Delta$ ($\Gamma \models_{F,t} \Delta$) is shorthand for $\Gamma \models_{F,i} \alpha$ for all $\alpha \in \Delta$ ($\Gamma \models_{F,i} \alpha$ for all $\alpha \in \Delta$).

Let Γ' be the inferential closure of Γ under the inference rule of Scope Instantiation. Let Γ'' be the subset of Γ that contains only $L_F(\mathbf{N})$ sentences that are relativised to an application α as opposed to a set A of applications. Clearly, (i) Γ, Γ' and Γ'' have the same $L_F(\mathbf{N})$ models in the context of $P_1(\mathbf{N})$. That is,

$$MOD(\Gamma) \cap MOD(P_1(\mathbf{N})) \cap MOD(D(\mathbf{N})) \cap EXT(\mathcal{A}_o, V_t(\mathbf{N}) \cup V_d(\mathbf{N})) =$$

$$MOD(\Gamma') \cap MOD(P_1(\mathbf{N})) \cap MOD(D(\mathbf{N})) \cap EXT(\mathcal{A}_o, V_t(\mathbf{N}) \cup V_d(\mathbf{N})) =$$

$$MOD(\Gamma'') \cap MOD(P_1(\mathbf{N})) \cap MOD(D(\mathbf{N})) \cap EXT(\mathcal{A}_o, V_t(\mathbf{N}) \cup V_d(\mathbf{N})).$$

Because of the semantic constraints of finite frame logic, we can characterise the admissible extensions of an application α (as defined by $MOD(\Gamma'') \cap MOD(P_1(\mathbf{N})) \cap MOD(D(\mathbf{N})) \cap EXT(\mathcal{A}_o, V_t(\mathbf{N}) \cup V_d(\mathbf{N}))$ by a set of Carnapian state descriptions, i.e., a set of a set of first-order literals. Any such set of literals represents uniquely an admissible extension of α. Let S_α be the set of state descriptions that represent the admissible extensions of α. Let γ_α be a sentence in disjunctive normal form such that any conjunction is a conjunction of all literals of a specific state description in S_α, and for any state description in S_α we have a corresponding conjunction that is a disjunct of γ_α. Clearly, (ii) $Cn(\gamma_\alpha) = Cn(S_\alpha)$, where Cn designates the consequence operation of classical logic.

(iii) Let Γ_α be the subset of Γ'' such that all members of Γ_α are relativised to application α. Further, let Γ'_α be the set of first-order sentences such that $\beta' \in \Gamma'_\alpha$ iff $\beta'|_\alpha \in \Gamma_\alpha$. Since the postulates $P_1(\mathbf{N})$ build upon the semantic rules of first-order logic and impose the additional constraint that all admissible extensions of an application α of $\mathbf{T}$ are models of $\mathbf{T}$, we have $Cn(\Gamma'_\alpha), \lambda_\alpha \models \gamma_\alpha$, where λ_α is the substantial law of the theory-element $\mathbf{T}$ of α. By completeness of first-order logic, this implies (iv): $Cn(\Gamma'_\alpha), \lambda_\alpha \vdash \gamma_\alpha$. Since all inference rules of first-order natural deduction are available in frame logic (relativised to an application α) as well as the inference rule Law Application 2, (i), (ii), (iii) and (iv) imply that $\Gamma \vdash_F \gamma_\alpha|_\alpha$, for any application α in $\mathbf{N}$. (Note that γ_α is a first-order sentence, while $\gamma_\alpha|_\alpha$ is an $L_F(\mathbf{N})$ sentence.) The union of the sets $\{\gamma_\alpha|_\alpha\}$ forms the set Δ such that $\Gamma \vdash_F \models_{F,i} \Delta$.

So far, we have taken only intramodular frame-logical consequences into account. Let us now consider the transmodular frame-logical consequences. Let the sets $\mathbf{S}'_a$ and $\mathbf{S}_a$ be defined as follows:

$$\mathbf{S}'_a = MOD(\Gamma) \cap MOD(P_1(\mathbf{N})) \cap MOD(D(\mathbf{N})) \cap EXT(\mathcal{A}_o, V_t(\mathbf{N}) \cup V_d(\mathbf{N}))$$

$$\mathbf{S}_a = MOD(\Gamma) \cap MOD(P_F(\mathbf{N})) \cap MOD(D(\mathbf{N})) \cap EXT(\mathcal{A}_o, V_t(\mathbf{N}) \cup V_d(\mathbf{N})).$$

Recall that (v) the set Δ (which is given by the union of the sets $\{\gamma_\alpha|_\alpha\}$) uniquely characterises the set $\mathbf{S}_a$ since the members of Δ uniquely characterise the admissible extensions of α, for any application α, in the context of the postulates $P_1(\mathbf{N})$. Since $P_F(\mathbf{N}) = P_1(\mathbf{N}) \cup P_2(\mathbf{N})$, we can obtain the set $\mathbf{S}_a$ from $\mathbf{S}'_\mathbf{a}$ by imposing the constraint of link satisfaction. This semantic constraint corresponds to the following syntactic operation on the sentence γ_α: for any applications α and β, and any literal l such that l is in the language of both α and β, if l is a conjunct in all disjuncts of γ_α, then we "delete" all disjuncts of γ_β of which l is not a conjunct. (If l is a conjunct in all disjuncts of γ_α, this means that l is verified by all admissible extensions of $\mathbf{I}(\alpha)$. So, one of the link postulates $P4.1(\mathbf{T})$ and $P4.2(\mathbf{T})$ applies.) Let Δ' be the set of sentences $\gamma'_\alpha|_\alpha$ that are obtained from the sentences $\gamma_\alpha|_\alpha$ by this operation, for all applications α in $\mathbf{N}$. Since there is a proof-theoretic representation of link satisfaction by Link Exploitation 1 and 2 in the context of the first-order natural deduction rules relativised to applications, it holds that (vi) $\Delta \vdash_F \delta$ for any $\delta \in \Delta'$. Because of (v) and since the present syntactic operation on the members of Δ amounts to imposing the constraint of link satisfaction, (vii) Δ' uniquely characterises the set $\mathbf{S}_a$ of $L_F(\mathbf{N})$ structures, i.e.,

$$\mathbf{S}_a = MOD(\Delta') \cap MOD(P_F(\mathbf{N})) \cap MOD(D(\mathbf{N})) \cap EXT(\mathcal{A}_o, V_t(\mathbf{N}) \cup V_d(\mathbf{N})).$$

Now, we have to distinguish two cases: (a) ϕ (for which $\Gamma \models_F \phi$) is an $L_F(\mathbf{N})$ sentence that is relativised to an application α, and (b) ϕ is relativised to a set A of applications. Suppose (a) so that ϕ is given by a sentence $\beta|_\alpha$. Consider the sentence $\gamma'_\alpha|_\alpha$ such that $\gamma'_\alpha|_\alpha \in \Delta'$. Note that the corresponding first-order sentence γ'_α is a syntactic characterisation of the admissible extensions of α, admissible in the context of $P_F(\mathbf{N})$. Because of (vii), $\gamma'_\alpha \models \beta$. Because of completeness of first-order logic, this implies $\gamma'_\alpha \vdash \beta$. Since all inference rules of first-order natural deduction are available in frame logic (relativised to an application α), we can infer from this that $\gamma'_\alpha|_\alpha \vdash_F \beta|_\alpha$. Using $\Gamma \vdash_F \delta$ for all $\delta \in \Delta$ and $\Delta \vdash_F \gamma'_\alpha|_\alpha$, this implies $\Gamma \vdash_F \phi$. This concludes the proof for the case (a). The proof for case (b) is analogous to that of (a), with the qualification that the inference rules Scope Comprehension and Scope Extension need to be exploited in the obvious way. □

Theorem 8.5 (Soundness) *Let Γ be a set of $L_{RF}(\mathbf{N})$ sentences and ϕ be such a sentence. If $\Gamma \vdash_{RF} \phi$, then $\Gamma \models_{RF} \phi$.*

Soundness can be proven in the standard way, i.e., by using an inductive proof on the number of inferences of a derivation. For the induction base, suppose the number of inferences is zero. Then, the derivation has the form $\{\phi\} \vdash_{RF} \phi$. The tree that consists of just a single sentence trivially satisfies the definition of a derivation (Definition 8.9). Furthermore, $\{\phi\} \models_{RF} \phi$ holds trivially if ϕ is an $L_{RF}(\mathbf{N})$ sentence.

Induction step. Let Γ' be the set Γ united with all singletons $\{\phi\}$ such that ϕ is a premise of a so far uncompleted subproof. Let n be the number of inferences drawn so far from sentences of Γ'. By the induction hypothesis, we know that (i) all so far derived sentences are logical consequences – in the sense of $\models_{RF}$ – of Γ'. Now, we need to show that (ii) any correct application of an inference rule leads to a sentence γ such that γ is a resolution-frame-logical consequence of the set of so far derived sentences, where the derivation is at step n. (i) and (ii) imply that γ is a resolution-frame-logical consequence of Γ'. So, we need to establish proposition (ii) for each inference rule of resolution frame logic.

For the inference rules Intramodular Resolution, Transmodular Resolution and Contradiction Introduction, the proof of proposition (ii) is straightforward and almost trivial. These three rules are instances of the standard, ground resolution inference rule, which is correct with respect to the semantics of classical propositional logic. The correctness of Literal Introduction 1 is trivial. To see the correctness of Literal Introduction 2 and Literal Introduction 3, recall that $\neg C_\alpha$ represents a negated disjunction of the form $\neg(l_1 \vee \ldots \vee l_n)$, where $C = \{l_1 \vee \ldots \vee l_n\}$ and for all i $(1 \leq i \leq n)$ $l_i \in L(V(\alpha))$. As is well known, such a disjunction is classically logically equivalent with $\neg l_1 \wedge \ldots \wedge \neg l_n$.

Showing the correctness of Negation Elimination 1 and 2 is a little bit more involved. Suppose □ has been derived within a subproof whose premise is $\neg C_\alpha$, where $C = \{l_1 \vee \ldots \vee l_n\}$. We need to show that, for all interpretations $\mathcal{V}(\mathbf{N})$, if $\mathcal{V}(\mathbf{N}) \models_{RF} \Gamma'$, then $\mathcal{V}(\mathbf{N}) \models_{RF} C_\alpha$. Suppose, for contradiction, there is an interpretation $\mathcal{V}(\mathbf{N})$ such that $\mathcal{V}(\mathbf{N}) \models_{RF} \Gamma'$ and $\mathcal{V}(\mathbf{N}) \not\models_{RF} C_\alpha$. By the semantics of clauses, this implies that $\mathcal{V}(\mathbf{N}) \models_{RF} \neg C_\alpha$. By $\neg C_\alpha \vdash_s \Box$ (which is the premise of Negation Elimination 1) and the induction hypothesis (all so far derived sentences are resolution-frame-logical consequences of Γ'), it follows that there is a literal l such that $\mathcal{V}(\mathbf{N}) \models_{RF} l_\alpha$ and $\mathcal{V}(\mathbf{N}) \models_{RF} \bar{l}_\beta$ (where $\alpha = \beta$ or $\alpha \neq \beta$). But this is impossible by the definition of $\mathcal{V}(\mathbf{N})$. Contradiction. Thus, we have established the correctness of Negation Elimination 1. The proof for Negation Elimination 2 is analogous. □

Theorem 8.6 (Completeness) *Let Γ be a set of $L_{RF}(\mathbf{N})$ sentences and ϕ be such a sentence. If $\Gamma \models_{RF} \phi$, then $\Gamma \vdash_{RF} \phi$.*

Lemma 8.1 *Let Γ be a set of $L_{RF}(\mathbf{N})$ sentences that are relative to one and the same application α. Let γ_α be an $L_{RF}(\mathbf{N})$ sentence. If $\Gamma_\alpha \models \gamma$, then $\Gamma \vdash_{RF} \gamma_\alpha$.*

This lemma says that, for sets of $L_{RF}(\mathbf{N})$ sentences that are relative to one and the same application, resolution frame logic is complete with respect to classical propositional logic. In the antecedent $\Gamma_\alpha \models \gamma$, $\models$ has its classical meaning, and clauses are taken to represent disjunctions of literals. Suppose $\Gamma_\alpha \models \gamma$. Hence, (i) $\Gamma_\alpha \cup \{\neg\gamma\}$ is classically unsatisfiable. (Recall that the sentences of Γ_α are not relativised to an application: $\Gamma_\alpha = \{\phi \mid \phi_\alpha \in \Gamma\}$.) Case (a): γ has the logical form of l (which implies that γ is a literal l). Case (b): γ has the logical form of C and $C = \{l_1, \ldots, l_n\}$ (which implies that γ is a clause C).

Case (a). By the resolution theorem of propositional logic (Theorem 8.4), (i) implies (ii) $\Gamma_\alpha \cup \{\bar{l}\} \vdash_R \Box$, where $\vdash_R$ stands for derivability in the standard ground resolution calculus, which has ground resolution as its only inference rule. (iii) The rule Intramodular Resolution is the ground resolution rule for sets of clauses that are relative to one and the same application. Since all members of Γ as well as ϕ are relative to a given α, (ii) and (iii) imply that $\Gamma \cup \{\bar{l}_\alpha\} \vdash_{RF} \Box$. By Literal Introduction 1, this implies $\Gamma, \bar{l}_\alpha \vdash_s \Box$. By Negation Elimination 2, we obtain $\Gamma \vdash_{RF} l_\alpha$. This concludes the proof for case (a).

Case (b): (i) implies (ii) $\Gamma_\alpha, \neg C \vdash_R \Box$. Using (iii), we can infer from this that $\Gamma, \neg C_\alpha \vdash_{RF} \Box$. Hence, $\neg C_\alpha \vdash_s \Box$, in the context of Γ. By Negation Elimination 1, this implies $\Gamma \vdash_{RF} C_\alpha$. □

Lemma 8.2 *Let Γ be a set of $L_{RF}(\mathbf{N})$ sentences. For all $\gamma \in \Gamma^*$, $\Gamma \vdash_{RF} \gamma$.*

Suppose $\gamma \in \Gamma^*$. We show by induction that (i), for all n ($n \in \mathbb{N}$) and for all $\Gamma^n \vdash_{RF} \gamma$, $\Gamma \vdash_{RF} \gamma$. The base case $n = 0$ holds true by definition of Γ^*. Induction step. Suppose $\gamma \in \Gamma^{n+1}$. We need to show that $\Gamma^n \vdash_{RF} \gamma$. Two cases need to be distinguished. Case (a): $\gamma \in \Gamma^n$. In this case, we are done with the induction step. Case (b): $\gamma \notin \Gamma^n$. By the definition of Γ^* and $\gamma \in \Gamma^{n+1}$, this implies that there is α such that $\Gamma^n_\alpha \models l$ and γ is l_β. Hence, by Lemma 8.1, monotonicity of $\vdash_{RF}$ and Γ^n_α being the set of formulas ϕ such that $\phi_\alpha \in \Gamma^n$, $\Gamma^n \vdash_{RF} l_\alpha$. Since l refers to a literal that is in the language of both $L(V(\alpha))$ and $L(V(\beta))$, this implies that we can derive l_β from Γ^n using Contradiction Introduction and Negation Elimination 2. Since γ is l_β, this implies that $\Gamma^n \vdash_{RF} \gamma$. Thus, we have established (i). The definition of Γ^* and (i) imply the claim of the lemma. □

We are now in a position to prove Theorem 8.6. Let Γ be a set of $L_{RF}(\mathbf{N})$ sentences and γ_α be such a sentence. Suppose $\Gamma \models_{RF} \gamma_\alpha$. Then, there is no interpretation $\mathcal{V}(\mathbf{N})$ such that $\mathcal{V}(\mathbf{N}) \models \Gamma$ and $\mathcal{V}(\mathbf{N}) \not\models \gamma_\alpha$. By the definition of $\mathcal{V}(\mathbf{N}) \models \Gamma$, this implies that there is no $\mathcal{V}(\mathbf{N})$ such that $\mathcal{V}(\mathbf{N})_\alpha \models \Gamma^*_\alpha$ and $\mathcal{V}(\mathbf{N}) \not\models \gamma_\alpha$. Hence, $\Gamma^*_\alpha \models \gamma$. By Lemma 8.1 and monotonicity of $\vdash_{RF}$, $\Gamma^* \vdash_{RF} \gamma_\alpha$. By Lemma 8.2, we can infer from this that $\Gamma \vdash_{RF} \gamma_\alpha$. □

Theorem 8.7 *Let Γ be a finite set of $L_{RF}(\mathbf{N})$ sentences. Further, suppose there is a constant c_1 such that, for all applications α in $\mathbf{N}$, $|V(\alpha)| \leq c_1$. The problem of determining whether Γ is satisfiable – in the sense of resolution frame logic – has* **P**-*time complexity.*

Lemma 8.3 *Let L be a set of literals. C and C' are clauses such that $|C| > 1$ and $|C'| > 1$. We further assume that $C' \subseteq C$. $\vdash_R$ stands for the relation of derivability in the standard, propositional resolution calculus. $\models$ designates the relation of logical consequence in classical logic. Then,*

(1) $L, C \vdash_R C'$ *iff* $L, \bigvee C \models \bigvee C'$
(2) $L, C \vdash_R \Box$ *iff* $L, \bigvee C \models \bot$

where $\Box$ designates the empty clause, which results from resolving complementary literals with one another. $\bigvee C$ designates some disjunction of the members of C if $|C| > 1$. If $|C| = 1$, $\bigvee C$ stands for the unique member of C.

Ad (1). Forward: suppose $L, C \vdash_R C'$. Then, for any literal $l \in C \setminus C', \bar{l} \in L$. Then, clearly, any propositional structure that satisfies $L \cup \{\bigvee C\}$ must also satisfy $\bigvee C'$. Hence, $L, \bigvee C \models \bigvee C'$.

Backward: $L, \bigvee C \models \bigvee C'$. Then, (i) for any literal $l \in C \setminus C', \bar{l} \in L$. For, otherwise, there would be a propositional structure that satisfies L, $\bigvee C$, but not $\bigvee C'$. (i) implies that $L, C \vdash_R C'$.

Ad (2). Forward: suppose $L, C \vdash_R \Box$. Then, for any literal $l \in C, \bar{l} \in L$. Hence, $L, \bigvee C \models \bot$.

Backward: suppose $L, \bigvee C \models \bot$. Then, for any literal $l \in C, \bar{l} \in L$. Hence, $L, C \vdash_R \Box$. □

We are now in a position to prove Theorem 8.7. Algorithm 8.1 has two parts. Part A determines intramodular inferences, while Part B determines the transmodular inferences. Soundness and completeness of Part A follow from Lemma 8.3 and the fact that standard, propositional Resolution remains valid for intramodular reasoning. Soundness and completeness of Part B follow from the definition of Γ^{n+1} within the inductive definition of Γ^*. Hence, (i) Algorithm 8.1 allows us to determine the satisfiability of a set Γ of $L_{RF}(\mathbf{N})$ sentences, provided clauses are represented by actual disjunctions of literals. By the proof of Theorem 8.3, we know that (ii) Algorithm 8.1 has **P**-time complexity. (i) and (ii) imply Theorem 8.7. □

Proposition 8.2 *Let all symbols and premises be as in Proposition 8.1. Let n be the number of applications in $\mathbf{N}$. m is the maximal cardinality of a clause C_α such that C_α is a member of an element in $\Theta_R(\mathbf{N})$. Then, the test for consistency of the set Γ by Algorithm 8.2 has the computational complexity of $O(n^2 \cdot m)$.*

Note that Algorithm 8.2 consists of two parts. Part A determines, for all α in $\mathbf{N}$, $C(\alpha)$ using only intramodular inferences. Part B (given by the repeat-until loop) further determines $C(\alpha)$ using transmodular inferences. The algorithm terminates if an inconsistency is detected, or the sets $C(\alpha)$ are determined such that no application can be further determined by an inference, or both. We yield the output *satisfiable* if and only if all sets $C(\alpha)$ are non-empty and no transmodular inference can be drawn any more.

Suppose we impose a total order of literals on the sets $C(\alpha)$ and $L_0(\alpha)$ such that the resulting two orders conform to one another. Then, we obtain for the computational complexity of Part A:

$$Comp(intra) = O(n \cdot m).$$

As for Part B, let us first determine the computational complexity of determining whether or not, for a given pair (α, β), $\overline{L(\alpha)} \subseteq C(\beta)$. (i) The complexity of this test is obviously linear in m. (ii) Within a single instance of the repeat-until loop the test for $\overline{L(\alpha)} \subseteq C(\beta)$ is carried out for at most $n - 1$ pairs (α, β) of applications, where α is fixed. Also, (iii) the number of repeat-until loops is bounded by n since any application cannot be "put" twice into the dynamic set $TR(\mathbf{N})$. (This feature of the algorithm is motivated by Lemma 8.4, which implies that – for all applications α – $L(\alpha)$ is empty or a singleton.) (i), (ii) and (iii) imply

$$Comp(trans) = O(n^2 \cdot m)$$

where $Comp(trans)$ stands for the complexity of determining transmodular inferences in Part B of the algorithm. Now, $Comp(intra) = O(n \cdot m)$ and $Comp(trans) = O(n^2 \cdot m)$ imply that the complexity of Algorithm 8.2 is $O(n^2 \cdot m)$. □

Proposition 8.3 *Let Γ be a consistent set of $L_{RF}(\mathbf{N})$ sentences. For any application α, there is exactly one $C_\alpha \in \Gamma$ such that $|C_\alpha| > 1$. Let $DA(\beta)$ be the set of applications α such that a sentence l_α determines – directly or indirectly – a literal of β, as a consequence of running Algorithm 8.2. Let $\Gamma_{DA(\beta)}$ be the corresponding set of sentences: $\Gamma_{DA(\beta)} = \{\gamma_\alpha \mid \alpha \in DA(\beta)\}$. Suppose there is l_ε such that $\varepsilon \notin DA(\beta)$ and $\Gamma_{DA(\beta)}, l_\varepsilon \vdash_{RF} \Box$. But for any $\alpha \in DA(\beta)$ such that $\alpha \neq \beta$, we have $\Gamma_{DA(\alpha)}, l_\varepsilon \nvdash_{RF} \Box$. Also, $\Gamma_\beta, l_\varepsilon \nvdash_{RF} \Box$. Then, for any $\Gamma' \subset \Gamma_{DA(\beta)}$, it holds that $\Gamma', l_\varepsilon \nvdash_{RF} \Box$.*

Let us first establish two lemmas, and then prove this proposition.

Lemma 8.4 *Let all symbols and premises be as in Proposition 8.3. Let $C_\beta \in \Gamma$ be a clause that represents a disjunction. Then, there is at most one literal l_β such that $\Gamma \vdash_{RF} l_\beta$ and $\Gamma \setminus \{C_\beta\} \nvdash_{RF} l_\beta$.*

This lemma says that C_β can be relevant for the derivation of at most one literal l_β. To be more precise, in the language $L(V(\beta))$ there can be at most one literal l such that l_β can be derived from Γ but not from $\Gamma \setminus \{C_\beta\}$. To see why this lemma holds, note that the premises of Proposition 8.3 assume that, for every non-composite application α, there is exactly one clause C_α in Γ such that $|C_\alpha| > 1$. Hence, every application of the inference rules Intramodular and Transmodular Resolution is such that a clause containing more than one literal is resolved with a literal. Hence, (i) all frame-logical resolutions from C_β – as well as from sentences inferred from C_β by Intramodular or Transmodular Resolution – are of the form

$$\frac{C'_\beta \quad \bar{l}_{i_\alpha}}{(C' - l_i)_\beta}$$

where $l_i \in C'$. α may or may not be identical with β. It is almost obvious that using a sequence of such inferences, we can infer at most one literal l_β from a clause C_β that represents a disjunction. Suppose, for contradiction, C_β is relevant for the derivation of two literals in the language of $L(V(\beta))$. That is, (ii) there are $l_{1\beta}$ and $l_{2\beta}$ such that $\Gamma \vdash_{RF} l_{1\beta}$, $\Gamma \vdash_{RF} l_{2\beta}$, while $\Gamma \backslash \{C_\beta\} \not\vdash_{RF} l_{1\beta}$ and $\Gamma \backslash \{C_\beta\} \not\vdash_{RF} l_{2\beta}$. Then, because of (i), (iii) there is a frame-logical derivation such that all members of C_β but $l_{1\beta}$ have been "eliminated" by applications of Intramodular or Transmodular Resolution. (A literal l is eliminated from a clause C if C is resolved with $\bar{l}$ such that the resolvent is $C \setminus \{l\} \subset C$.) Further, (ii) implies (iv) there is a frame-logical derivation such that all literals of C_β but $l_{2\beta}$ have been "eliminated". From this we can infer that $l_{1\beta}$ can be derived from $\Gamma \backslash \{C_\beta\}$. This conclusion, however, contradicts (ii). □

The proof of Lemma 8.4 exploits the observation that – when assuming the premises of Proposition 8.3 – all frame-logical resolutions are such that a disjunction is resolved with a literal. This observation implies another lemma:

Lemma 8.5 *Let L be a set of literals of the logical form l_β. C_β is a clause that represents a disjunction. $L, C \vdash_{RF} l'_\beta$. Further, assume that, for any $L' \subset L$, we have $L', C_\beta \not\vdash_{RF} l'_\beta$. Then, there is exactly one literal $l \in C$ such that $\bar{l}_\beta \notin L$.*

In other words, all but one literal in C_β must be determinable to be false – by resolution with members of L – in order for C_β to be relevant for the derivation of a literal l'_β in the context of L.

Let us now resume the proof of Proposition 8.3. Let $DA(\beta)$ and $\Gamma_{DA(\beta)}$ be as explained in this proposition. Suppose $\Gamma_{DA(\beta)}$ is frame-logically inconsistent with some l_ε. Hence, there is l_β such that $\Gamma_{DA(\beta)} \vdash_{RF} \bar{l}_\beta$. Let γ_α be some arbitrary member of $\Gamma_{DA(\beta)}$. Then, there is a sequence $\alpha_1, \ldots, \alpha_m$ ($m \geq 0$) of applications such that α determines some literal of α_1, α_1 some literal of α_2, etc., and α_m determines a literal l'_β (which may or may not be identical with l_β). By Lemmas 8.4 and 8.5, we can infer that C_α is essential for the determination of the literal determined in α_1. To be more precise, there is a literal l_{α_1} such that $\Gamma_{DA(\beta)} \vdash_{RF} l_{\alpha_1}$ and $\Gamma_{DA(\beta)} \setminus \{C_\alpha\} \not\vdash_{RF} l_{\alpha_1}$. By iteration of the last step of reasoning, we infer that there is a literal l'_β such that $\Gamma_{DA(\beta)} \vdash_{RF} l'_\beta$ and $\Gamma_{DA(\beta)} \setminus \{C_\alpha\} \not\vdash_{RF} l'_\beta$. By Lemma 8.5, we can infer from this that $\Gamma_{DA(\beta)} \vdash_{RF} l_\beta$ and $\Gamma_{DA(\beta)} \setminus \{C_\alpha\} \not\vdash_{RF} l_\beta$. Because of this and since – for any $\alpha \in DA(\beta)$ such that $\alpha \neq \beta$ – we have

$\Gamma_{DA(\alpha)}, l_\varepsilon \not\vdash_{RF} \Box$, it holds that $\Gamma_{DA(\beta)} \setminus \{C_\alpha\}$ is consistent with l_ε. This concludes the proof of Proposition 8.3. $\Box$

Theorem 8.10 *Let $\Theta_R(\mathbf{N})$ be a structuralist default theory in resolution frame logic. For every non-composite application α, there is exactly one clause C_α in $\Theta_R(\mathbf{N})$ such that $|C_\alpha| > 1$. Further, any composite application is non-vacuous in the sense of Explanation 8.3. Then, TMS$_R$ infers a literal γ from $(\Theta_R(\mathbf{N}) \oplus_i \phi)$ if and only if $B_{\vdash_{RF}}(\Theta_R(\mathbf{N}) \oplus_i \phi) \models_{RF} \gamma$.*

To prove this theorem, we first need to show several lemmas.

Lemma 8.6 *Let Γ be a set of $L_{RF}(\mathbf{N})$ sentences each of which is relative to an application $\alpha \in H_s(\mathbf{N})$. For any non-composite application α, there is exactly one $C_\alpha \in \Gamma$ such that $|C_\alpha| > 1$. Further, suppose $\Gamma \not\models_{RF} \Box$. Then, TMS$_R$ infers – in the sense of Explanation 8.4 – an $L_{RF}(\mathbf{N})$ literal γ if and only if $\Gamma \models_{RF} \gamma$.*

(a) Let us first suppose that there are no composite applications. In resolution frame logic, we distinguish between intramodular and transmodular inferences. Hence, in order to show the claim of the lemma, we need to show that TMS$_R$ captures completely and correctly all intramodular and transmodular inferences. By the assumptions of the lemma, the information given by a modular unit $\alpha \in H(\mathbf{N})$ of reasoning is given by a clause C_α and literals l_α. Hence, Lemma 8.3 applies to the set Γ_α, where $\Gamma_\alpha = \{\phi \mid \phi_\alpha \in \Gamma\}$. Part A of Algorithm 8.2 (which is used in TMS$_R$), therefore, correctly and completely determines all intramodular inferences from Γ_α. Moreover, it is obvious that Part B of Algorithm 8.2 (which is used by TMS$_R$ as well) correctly and completely determines all transmodular inferences. Hence, TMS$_R$ infers an $L_{RF}(\mathbf{N})$ literal γ if and only if $\Gamma \models_{RF} \gamma$.

(b) Let us now lift the assumption that there are no composite applications. The reasoning of (a) for transmodular inferences remains in place as it is. We have assumed that, in the presence of composite applications, intramodular inferences are determined by a standard resolution algorithm, which is refutation complete. Using such an algorithm, TMS$_R$ determines, for all α such that $\Gamma_\alpha \neq \emptyset$, which literals are entailed by Γ_α. Hence, TMS$_R$ infers an $L_{RF}(\mathbf{N})$ literal γ if and only if $\Gamma \models_{RF} \gamma$. $\Box$

Lemma 8.7 *Let Γ be as in Lemma 8.6 with the following qualifications. Let A be a set of applications α such that $\Gamma_\alpha \neq \emptyset$. ($\Gamma_\alpha = \{\phi \mid \phi_\alpha \in \Gamma\}$.) There is at most one application $\alpha \in A$ such that $\alpha \in H_d(\mathbf{N})$. All but possibly one application of A are in $H_s(\mathbf{N})$. Then, TMS$_R$ detects an inconsistency of Γ if and only if $\Gamma \models_{RF} \Box$.*

The proof of this lemma is closely analogous to the proof of Lemma 8.6. Let us describe the proof idea. (i) For non-composite applications α, Part A of Algorithm 8.2 (which is used by TMS$_R$) correctly and completely determines all intramodular inferences from Γ_α. (ii) Part B of Algorithm 8.2 (which is used by TMS$_R$ as well) correctly and completely determines all transmodular inferences. (iii) For composite applications α, a standard resolution algorithm correctly and completely determines all intramodular inferences if the inferred sentence is a literal l_α. (iv) By completeness and soundness of resolution frame logic, $\Gamma \models_{RF} \Box$ iff there are literals l_α and l'_β such that $\Gamma \vdash_{RF} l_\alpha$, $\Gamma \vdash_{RF} l'_\beta$ and $\{l_\alpha, l'_\beta\} \vdash_{RF} \Box$, where α may or may not be identical with β. (i), (ii), (iii) and (iv) imply that TMS_R detects an inconsistency iff $\Gamma \models_{RF} \Box$. □

Lemma 8.8 *Let $\Theta_R(\mathbf{N})$ be a structuralist default theory in resolution frame logic. For every non-composite application α, there is exactly one clause C_α in $\Theta_R(\mathbf{N})$ such that $|C_\alpha| > 1$. Further, any composite application is non-vacuous in the sense of Explanation 8.3. Then, TMS_R determines the sets $H_s(\mathbf{N})$, $H_d(\mathbf{N})$ and $H_u(\mathbf{N})$ according to Definition 8.13. That is, if ϕ_α is an $L_{RF}(\mathbf{N})$ sentence, then*

(1) $\phi_\alpha \in B_{\vdash_{RF}}(\Theta_R(\mathbf{N}))$ *iff* $\alpha \in H_s(\mathbf{N})$
(2) $\rceil\phi_\alpha \in D_{\vdash_{RF}}(\Theta_R(\mathbf{N}))$ *iff* $\alpha \in H_d(\mathbf{N})$
(3) $\phi_\alpha \in R_{\vdash_{RF}}(\Theta_R(\mathbf{N}))$ *iff* $\alpha \in H_u(\mathbf{N})$.

We prove this lemma by induction on the length of the sequence S of revision formulas. For the induction base, we consider the revision of the empty belief base with a single $L_{RF}(\mathbf{N})$ sentence ϕ_α. Note that α cannot be a composite application since any composition must be based on other applications. Hence, ϕ is a clause or a literal. Obviously, the claim of the lemma is trivial in this case. Let us call a set A of applications consistent – relative to Γ – iff $\{\phi_\gamma \mid \phi_\gamma \in \Gamma \text{ and } \gamma \in A\} \not\models_{RF} \Box$.

Induction step. Let $\Theta_R(\mathbf{N}')$ be given by $\Theta_R(\mathbf{N}) \oplus_j \phi_\alpha$, where j is the epistemic rank of $\mathbf{T}(\alpha)$. ϕ_α is at position $i+1$ of sequence S.

(a) Let us first assume that the sequence of revision formulas $\alpha_1, \ldots \alpha_k$ accords with the epistemic ordering of this sequence, i.e., $i < j$ $(i, j \leq k)$ implies that $\alpha_i \leq \alpha_j$. Also, let us suppose that there are no composite applications.

Ad (1). Forward: suppose (i) $\phi_\alpha \in B_{\vdash_{RF}}(\Theta_R(\mathbf{N}'))$. By the induction hypothesis, (ii) propositions (1)–(3) of the lemma hold for $\Theta_R(\mathbf{N})$. Recall that (iii) ϕ_α is classified as successful iff α passes the consistency test with all sets $H_s(\mathbf{N}) \cup \{\delta\}$ such that $\delta \in H_d(\mathbf{N})$. By Lemmas 8.6 and 8.7 and the general definition of $B_{\vdash_{RF}}(\Theta_R(\mathbf{N}))$, (i), (ii) and (iii) imply that α passes this consistency test. Hence, $\alpha \in H_s(\mathbf{N}')$.

Backward: suppose (i) $\alpha \in H_s(\mathbf{N}')$. Again, by the induction hypothesis, (ii) Propositions (1) and (2) of the lemma hold for $\Theta_R(\mathbf{N})$. (i) implies that α passed the consistency test with all sets $H_s(\mathbf{N}) \cup \{\delta\}$ such that $\delta \in H_d(\mathbf{N})$. Using (ii), Lemmas 8.6 and 8.7, and the general definition of $B_{\vdash_{RF}}(\Theta_R(\mathbf{N}))$, we can infer from this that $\phi_\alpha \in B_{\vdash_{RF}}(\Theta_R(\mathbf{N}'))$. Thus, we have shown both directions of Proposition (1) of the lemma.

Ad (3). Forward: suppose (i) $\phi_\alpha \in R_{\vdash_{RF}}(\Theta_R(\mathbf{N}'))$. By the induction hypothesis, (ii) Proposition (1) of the lemma holds for $\Theta_R(\mathbf{N})$. Recall that (iii) α is classified as unsuccessful iff α does not pass the consistency test with the set $\{\beta \in H_s(\mathbf{N}) \mid \beta < \alpha\}$. By Lemmas 8.6 and 8.7 and the general definition of $R_{\vdash_{RF}}(\Theta_R(\mathbf{N}'))$, (i), (ii) and (iii) imply that α does not pass this consistency test. Hence, $\alpha \in H_u(\mathbf{N}')$.

Backward: suppose (i) $\alpha \in H_u(\mathbf{N}')$. Again, by the induction hypothesis, (ii) Proposition (1) of the lemma holds for $\Theta_R(\mathbf{N})$. (i) implies that α does not pass the consistency test with the set $\{\beta \in H_s(\mathbf{N}) \mid \beta < \alpha\}$. Using (ii), Lemmas 8.6

and 8.7, and the general definition of $R_{\vdash_{RF}}(\Theta_R(\mathbf{N}'))$, we can infer from this that $\alpha \in R_{\vdash_{RF}}(\Theta_R(\mathbf{N}'))$. Thus, we have shown both directions of Proposition (3) of the lemma.

Ad (2). By propositions (1) and (3), we know that the revision formula ϕ_α is classified correctly. That is, $\phi_\alpha \in D_{\vdash_{RF}}(\Theta_R(\mathbf{N}'))$ iff $\alpha \in H_d(\mathbf{N}')$. However, in order to show Proposition (2), we also need to show that $\beta \in H_s(\mathbf{N})$ is doubted iff all $\psi_\beta \in D_{\vdash_{RF}}(\Theta_R(\mathbf{N}'))$. To be more precise, we need to show that $(*)$ $\beta \in H_s(\mathbf{N})$ and $\beta \in H_d(\mathbf{N}')$ iff, for all $\psi_\beta \in B_{\vdash_{RF}}(\Theta_R(\mathbf{N}))$, we have $\psi_\beta \in D_{\vdash_{RF}}(\Theta_R(\mathbf{N}'))$. To prove proposition $(*)$, it is helpful to introduce some notational conventions. Where B is a set of applications and Γ the set of all $L_{RF}(\mathbf{N})$ sentences in $\Theta_R(\mathbf{N})$, $\Gamma_B = \{\psi \mid \psi_\gamma \in \Gamma \text{ and } \gamma \in B\}$. Likewise, where Γ is the set of all $L_{RF}(\mathbf{N})$ sentences in $\Theta_R(\mathbf{N})$, we use $\Gamma'_\alpha = \{\phi_\alpha \mid \phi_\alpha \in \Gamma\}$. (Note that $\Gamma'_\alpha \neq \Gamma_\alpha = \{\phi \mid \phi_\alpha \in \Gamma\}$.) Finally, recall that we call a set A of applications consistent – relative to Γ – iff $\{\phi_\gamma \mid \phi_\gamma \in \Gamma \text{ and } \gamma \in A\} \not\models_{RF} \Box$.

Forward: suppose $\beta \in H_s(\mathbf{N})$ and $\beta \in H_d(\mathbf{N}')$. Hence, the revision formula α must have been doubted by some application $\gamma \in H_s(\mathbf{N}) \cup H_d(\mathbf{N})$ such that β determines α or γ, or both. Let D be the set of applications that determine α or γ, or both. The TMS$_R$ algorithm is such that (i) the set $(D)_{=\phi_\alpha}$ of applications is doubted. Hence, (ii) $\beta \in (D)_{=\phi_\alpha}$. By Proposition 8.3, we know that Γ_D is minimally inconsistent in $\Gamma_{(D)_{=\phi_\alpha}}$. By the induction hypothesis, we know that $\Gamma_{(D)_{=\phi_\alpha}} \subset B_{\vdash_{RF}}(\Theta_R(\mathbf{N}))$. Hence, there is a choice function such that, for all $\psi_\beta \in B_{\vdash_{RF}}(\Theta_R(\mathbf{N}))$, we have $\psi_\beta \in D_{\vdash_{RF}}(\Theta_R(\mathbf{N}'))$. Put differently, there is a choice function such that $\Gamma'_\beta \subseteq B_{\vdash_{RF}}(\Theta_R(\mathbf{N}))$, while $\Gamma'_\beta \subseteq D_{\vdash_{RF}}(\Theta_R(\mathbf{N}'))$. One such choice function suffices, thanks to adopting the epsilon semantics of local truth, in the definition of the TM belief revision scheme. This concludes the forward direction of $(*)$.

Backward: suppose for all $\psi_\beta \in B_{\vdash_{RF}}(\Theta_R(\mathbf{N}))$, we have $\psi_\beta \in D_{\vdash_{RF}}(\Theta_R(\mathbf{N}'))$, where β is fixed. Hence, $\phi_\alpha \in D_{\vdash_{RF}}(\Theta_R(\mathbf{N}'))$, where ϕ_α is the revision formula. (Otherwise, we would not have $\psi_\beta \in D_{\vdash_{RF}}(\Theta_R(\mathbf{N}'))$.) By the forward direction of $(*)$, we know that there is a sequence S of revision formulas such that (i) α is found inconsistent with some application γ such that β determines α or γ, or both, where $\Gamma'_\gamma \subseteq B_{\vdash_{RF}}(\Theta_R(\mathbf{N})) \cup D_{\vdash_{RF}}(\Theta_R(\mathbf{N}))$, $\Gamma'_\beta \subseteq B_{\vdash_{RF}}(\Theta_R(\mathbf{N}))$ and the epistemic rank of β equals that of α. Thanks to the epsilon semantics of local truth (adopted in the TM belief revision scheme), we are free to assume that $\varepsilon x(x \in \mathbf{S}) = S$. Then, by the induction hypothesis, (i) implies $\gamma \in H_s(\mathbf{N}) \cup H_d(\mathbf{N})$ and $\beta \in H_s(\mathbf{N})$. Moreover, $\varepsilon x(x \in \mathbf{S}) = S$ and (i) imply that $\beta \in H_d(\mathbf{N}')$. Hence, $\beta \in H_s(\mathbf{N})$ and $\beta \in H_d(\mathbf{N})$. This concludes the proof of $(*)$.

Note finally, that the sequence S of revision formulas – by assumption – respects the epistemic ranking of applications (in the sense that it starts with those of the highest rank). Hence, TMS$_R$ will never undoubt or unreject an application. Hence, we do not have to consider the case that an application is moved from $H_d(\mathbf{N})$ to $H_s(\mathbf{N})$, or from $H_u(\mathbf{N})$ to $H_s(\mathbf{N})$. Thus, all possible additions to and retractions from the dynamic sets $H_s(\mathbf{N})$, $H_d(\mathbf{N})$ and $H_u(\mathbf{N})$ have been considered and shown to conform to Definition 8.13. This concludes the proof of the lemma on the above specified assumptions, i.e., no composite applications and the sequence S respects the epistemic ranking of applications.

(b) Let us now lift the assumption that there are no composite applications, without however lifting the assumption that the sequence of revision formulas $\alpha_1, \ldots \alpha_k$ accords with the epistemic ordering of this sequence, i.e., $i < j$ $(i, j \leq k)$ implies that $\alpha_i \leq \alpha_j$. The reasoning in (a) for propositions (1) and (3) of the lemma is not affected by the presence of composite applications. Let us therefore focus on proposition (2). By propositions (1) and (3), we know that the revision formula ϕ_α is classified correctly. That is, $\phi_\alpha \in D_{\vdash_{RF}}(\Theta_R(\mathbf{N}'))$ iff $\alpha \in H_d(\mathbf{N}')$. It remains to show that $(*)$ $\beta \in H_s(\mathbf{N})$ and $\beta \in H_d(\mathbf{N}')$ iff, for all $\psi_\beta \in B_{\vdash_{RF}}(\Theta_R(\mathbf{N}))$, we have $\psi_\beta \in D_{\vdash_{RF}}(\Theta_R(\mathbf{N}'))$.

Forward: the reasoning in (a) remains valid, with the following qualification. Proposition 8.3 does not suffice to infer that Γ_D is minimally inconsistent in $\Gamma_{(D)=\phi_\alpha}$. To obtain this conclusion, we also need the assumption that all composite applications are non-vacuous, which is in fact an assumption of Lemma 8.8.

Backward: this direction of the proof of $(*)$ in (a) remains valid without further modifications.

(c) Let us finally lift the assumption that the sequence of revision formulas $\alpha_1, \ldots \alpha_k$ accords with the epistemic ordering of this sequence, i.e., $i < j$ $(i, j \leq k)$ implies that $\alpha_i \leq \alpha_j$. Again, we need to distinguish two cases: (i) no application in $H_s(\mathbf{N}) \cup H_d(\mathbf{N})$ is rejected. (ii) The revision formula or some application in $H_s(\mathbf{N}) \cup H_d(\mathbf{N})$ is rejected. In the case of (i), the entire reasoning in (b) remains valid. Let us now deal with case (ii) and let β be an application that is rejected. Then, first, all applications γ that are determined by β (directly or indirectly) will be undetermined as regards the – direct or indirect – determinations of γ. Second, all such undetermined applications γ are "removed" from $H_d(\mathbf{N})$ or $H_u(\mathbf{N})$ (if they are a member of $H_d(\mathbf{N})$ or $H_u(\mathbf{N})$) and put into the ordered set $TR^<(\mathbf{N})$. Likewise, for all applications in $B(\beta)$, the set of applications blocked by β. Third, $H_s(\mathbf{N})$, $H_d(\mathbf{N})$ and $H_u(\mathbf{N})$ are revised by a sequence S' of these undetermined and unblocked applications γ, where the order of the sequence conforms to the epistemic priority of its members. Thus, (i) the reasoning of (b) applies to the revisions according to sequence S'. Since we adopted the semantics of local truth for the epsilon terms in the TM belief revision scheme, (ii) we can choose S' as the sequence picked out of $\mathbf{S}$ by the epsilon terms in Definition 8.13. (iii) Propositions (1) to (3) have been shown on the assumptions of (b). (i), (ii) and (iii) imply the claim of Lemma 8.8. □

Lemmas 8.6 and 8.8 imply Theorem 8.10. □

Theorem 8.12 *Let $\Theta_R(\mathbf{N})$ be a structuralist default theory in resolution frame logic. ϕ is an $L_{RF}(\mathbf{N})$ sentence. For every non-composite application α, there is exactly one clause C_α in $\Theta_R(\mathbf{N})$ such that $|C_\alpha| > 1$. There is a constant c_1 such that, for any clause C_α that is a member of an element in $\Theta_R(\mathbf{N})$, $|C_\alpha| < c_1$. Any composite application is non-vacuous in the sense of Explanation 8.3. Further, the composition of applications in $\mathbf{N}$ is bounded by a constant c_2, in the sense that any composite application does not comprise more than c_2 applications. Then, the computational complexity of determining $B_{\vdash_{RF}}(\Theta_R(\mathbf{N}) \oplus_i \phi)$ is $O(n^2)$.*

Let n be the number of applications for all theory-elements $\mathbf{T} \in \mathbf{N}$. Then, a composition of c applications uniquely corresponds to a combination of c elements out of a set of n elements without repetitions. Hence, we have $(n!) \div ((n - c)!c!)$ different composite applications of cardinality c. Hence, (i) the number of applications – composite applications up to cardinality c and non-composite applications – is bounded by n^{c_2}, where c_2 is the maximal cardinality of a composite application. Recall that determining the set of literals that are inferences from a set D of disjunctions of literals is exponential in the cardinality of D. Since any composite application is bounded in size by c_2 and since, for every non-composite application α, there is exactly one clause C_α, this implies that determining the inferences within a composite application is bounded by a constant k^{c_2}. In the context of Theorem 8.11, (i) and (ii) imply that the additional computational costs ensued by compositions remain polynomial as long as there is an upper bound on the number of applications that make up a composition. □

Bibliography

1. Alchourrón, M. A., Gärdenfors, P., & Makinson, D. (1985). On the logic of theory change: Partial meet contraction functions and their associated revision functions. *Journal of Symbolic Logic, 50*, 510–530.
2. Alechina, N., Jago, M., & Logan, B. (2006). Resource-bounded belief revision and contraction. In M. Baldoni, U. Endriss, A. Omicini, & P. Torroni (Eds.), *Declarative agent languages and technologies III. Third international workshop, DALT 2005* (LNCS, Vol. 3904, pp. 141–154). Berlin/Heidelberg: Springer.
3. Amir, E., & McIlraith, S. (2005). Partition-based logical reasoning for first-order and propositional theories. *Artificial Intelligence, 162*(1–2), 49–88.
4. Andreas, H. (2008). Zur semantischen Analyse von Erscheinungssätzen bei Sellars. *Facta Philosophica, 10*(1), 269–283.
5. Andreas, H. (2010). A modal view of the semantics of theoretical sentences. *Synthese, 174*(3), 367–383.
6. Andreas, H. (2010). New account of empirical claims in structuralism. *Synthese, 176*(3), 311–332.
7. Andreas, H. (2011). A structuralist theory of belief revision. *Journal of Logic, Language, and Information, 20*(2), 205–232.
8. Andreas, H. (2012). Deductive reasoning in the structuralist approach. *Studia Logica, 101*(5), 1093–1113.
9. Andreas, H. (2014). Carnapian structuralism. *Erkenntnis, 79*(8), 1373–1391.
10. Andreas, H. (2014). On frames and theory-elements of structuralism. In R. Hagengruber & U. V. Riss (Eds.), *Philosophy, computing, and information science*. London: Routledge.
11. Andreas, H. (2018). Modular semantics for theories: An approach to paraconsistent reasoning. *Journal of Philosophical Logic, 47*(5), 877–912.
12. Andreas, H., & Schiemer, G. (2016). A choice-semantical approach to theoretical truth. *Studies in History and Philosophy of Science Part A, 58*, 1–8.
13. Antoniou, G. (1997). *Nonmonotonic reasoning*. Cambridge, MA: MIT Press.
14. Arora, S., & Barak, B. (2009). *Computational complexity: A modern approach*. Cambridge: Cambridge University Press.
15. Asser, G. (1957). Theorie der logischen Auswahlfunktionen. *Zeitschrift für mathematische Logik und Grundlagen der Mathematik, 3*, 30–68.
16. Baader, F., Calvanese, D., McGuinness, D. L., Nardi, D., & Patel-Schneider, P. F. (Eds.). (2003). *The description logic handbook: Theory, implementation, and applications*. New York: Cambridge University Press.

H. Andreas, *Dynamic Tractable Reasoning*, Synthese Library 420,
https://doi.org/10.1007/978-3-030-36233-1

17. Baader, F., Calvanese, D., McGuinness, D. L., Nardi, D., & Patel-Schneider, P. F. (2007). *The description logic handbook*. New York: Cambridge University Press.
18. Balzer, W., Moulines, C. U., & Sneed, J. (1987). *An architectonic for science. The structuralist program*. Dordrecht: D. Reidel Publishing Company.
19. Bartlett, F. C. (1967). *Remembering*. Cambridge: Cambridge University Press.
20. Bermúdez, J. L. (2010). *Cognitive science*. Cambridge: Cambridge University Press.
21. Bourbaki, N. (1968). *Elements of mathematics: Theory of sets*. Reading: Addison-Wesley.
22. Brachman, R. J., & Levesque, H. (2004). *Knowledge representation and reasoning*. Amsterdam: Elsevier.
23. Brewka, G. (1989). Preferred subtheories: An extended logical framework for default reasoning. In *Proceedings of Eleventh International Joint Conference on Artificial Intelligence, Detroit* (pp. 1043–1048). San Mateo, CA: Morgan Kaufmann.
24. Brewka, G. (1991). Belief revision in a framework for default reasoning. In *Proceedings of the Workshop on The Logic of Theory Change* (pp. 602–622). London: Springer.
25. Brewka, G., Dix, J., & Konolige, K. (1997). *Nonmonotonic reasoning. An overview*. Stanford: CSLI Publications.
26. Bruce, K. B. (2002). *Foundations of object-oriented languages: Types and semantics*. Cambridge, MA: MIT Press.
27. Cadoli, M. (1995). *Tractable reasoning in artificial intelligence*. Secaucus, NJ: Springer-Verlag New York, Inc.
28. Cadoli, M., & Schaerf, M. (1993). A survey of complexity results for non-monotonic logics. *Journal of Logic Programming, 17*, 127–160.
29. Calvanese, D., De Giacomo, G., Lembo, D., Lenzerini, M., & Rosati, R. (2007). Tractable reasoning and efficient query answering in description logics: The DL-Lite family. *Journal of Automated Reasoning, 39*(3), 385–429.
30. Carnap, R. (1937). *The logical syntax of language*. London: Routledge.
31. Carnap, R. (1939). *Foundations of logic and mathematics*. Chicago: University of Chicago Press.
32. Carnap, R. (1942). *Introduction to semantics*. Cambridge, MA: Harvard University Press.
33. Carnap, R. (1958). Beobachtungssprache und theoretische Sprache. *Dialectica, 12*, 236–248.
34. Carnap, R. (1961). On the use of Hilbert's ϵ-operator in scientific theories. In Y. Bar-Hillel, E. I. J. Poznanski, M. O. Rabin, & A. Robinson (Eds.), *Foundations of mathematics* (pp. 156–164). Jerusalem: Magnes Press.
35. Carruthers, P. (2006). *The architecture of the mind: Massive modularity and the flexibility of thought*. Oxford: Oxford University Press.
36. Cherniak, C. (1986). *Minimal rationality*. Cambridge, MA: MIT Press.
37. Chomsky, N. (1965). *Aspects of the theory of syntax*. Cambridge, MA: MIT Press.
38. Church, A. (1956). *Introduction to mathematical logic*. Princeton: Princeton University Press.
39. Churchland, P. (1987). Epistemology in the age of neuroscience. *Journal of Philosophy, 84*, 544–453.
40. Cook, S., & Reckhow, R. (1974). On the lengths of proofs in the propositional calculus (preliminary version). In *Proceedings of the Sixth Annual ACM Symposium on Theory of Computing* (pp. 135–148). New York: ACM.
41. Cosmides, L., & Tooby, J. (1994). Origins of domain specificity: The evolution of functional organization. In L. A. Hirschfeld & S. A. Gelman (Eds.), *Mapping the mind: Domain specificity in cognition and culture* (pp. 85–116). Cambridge: Cambridge University Press.
42. Dalal, M. (1988). Investigations into a theory of knowledge base revisions: Preliminary report. In *Proceedings of the Seventh National Conference on Artificial Intelligence (AAAI-88)*, St. Paul (pp. 475–479).
43. Davis, M., Logemann, G., & Loveland, D. (1962). A machine program for theorem-proving. *Communications of the ACM, 5*(7), 394–397.
44. Davis, M., & Putnam, H. (1960). A computing procedure for quantification theory. *Journal of the ACM, 7*(3), 201–215.

45. de Nivelle, H. (2005). Translation of resolution proofs into short first-order proofs without choice axioms. *Information and Computation, 199*(1), 24–54.
46. Ditmarsch, H. P. V. (2005). Prolegomena to dynamic logic for belief revision. *Synthese, 147*(2), 229–275.
47. Doyle, J. (1979). A truth maintenance system. *Artificial Intelligence, 12*, 231–271.
48. Doyle, J. (1979). A truth maintenance system. *MIT AI Lab Memo, 521*. http://hdl.handle.net/1721.1/5733
49. Doyle, J. (1992). Reason maintenance and belief revision. In P. Gärdenfors (Ed.), *Belief revision* (pp. 29–51). Cambridge: Cambridge University Press.
50. Ebbinghaus, H., & Flum, J. (1999). *Finite Model Theory*. Berlin: Springer.
51. Eckel, B. (2000). *Thinking in C++* (Vol. 1). Upper Saddle River, NJ: Prentice Hall.
52. Eckel, B. (2006). *Thinking in Java*. Upper Saddle River, NJ: Prentice Hall.
53. Fernández, M. (2004). *Programming languages and operational semantics* (Texts in computing, Vol. 1). London: King's College Publications.
54. Flum, J., & Grohe, M. (2006). *Parameterized complexity theory*. Berlin/Heidelberg: Springer.
55. Fodor, J. (1975). *The Language of thought*. New York: Crowell.
56. Fodor, J. (1983). *The Modularity of mind*. Cambridge, MA: MIT Press.
57. Fodor, J. (1987). Modules, frames, fridgeons, sleeping dogs, and the music of the spheres. In Z. W. Pylyshyn (Ed.), *The Robot's dilemma: The frame problem in artificial intelligence* (pp. 139–149). Norwood, NJ: Ablex.
58. Fodor, J. (2001). *The mind doesn't work that way*. Cambridge, MA: MIT Press.
59. Fodor, J. (2008). *LOT 2*. Oxford: Oxford University Press.
60. Frege, G. (1918/1919). Der Gedanke. Eine logische Untersuchung. *Beiträge zur Philosophie des deutschen Idealismus, 1*, 58–77.
61. Gärdenfors, P. (1988). *Knowledge in flux*. Cambridge, MA: MIT Press.
62. Gärdenfors, P., & Makinson, D. (1994). Nonmonotonic inference based on expectations. *Artificial Intelligence, 65*, 197–245.
63. Genesereth, M., & Kao, E. (2012). *Introduction to logic* (Synthesis Lectures on Computer Science). San Rafael: Morgan & Claypool. https://doi.org/10.2200/S00432ED1V01Y201207CSL005.
64. Gigerenzer, G., & Selten, R. (Eds.). (2001). *Bounded rationality: The adaptive toolbox*. Cambridge: MIT Press.
65. Gigerenzer, G., & Todd, P. M. (1999). *Simple heuristics that make us smart*. New York: Oxford University Press.
66. Goldman, A. (1979). What is Justified Belief? In G. S. Pappas (Ed.), *Justification and knowledge* (pp. 1–23). Dordrecht: Reidel.
67. Grove, A. (1988). Two modellings for theory change. *Journal of Philosophical Logic, 17*, 157–170.
68. Hansson, S. O. (1993). Reversing the Levi identity. *Journal of Philosophical Logic, 22*(6), 637–669.
69. Hansson, S. O. (1999). *A textbook of belief dynamics. Theory change and database updating*. Dordrecht: Kluwer.
70. Hansson, S. O. (2003). Ten philosophical problems in belief revision. *Journal of Logic and Computation, 13*, 37–49.
71. Hayes, P. (1985). The logic of frames. In B. L. Webber & N. J. Nilsson (Eds.), *Readings in Artificial Intelligence* (pp. 451–458). San Francisco: Morgan Kaufmann.
72. Hayes, P. (1987). What the frame problem is and isn't. In Z. W. Pylyshyn (Ed.), *The Robot's dilemma: The frame problem in artificial intelligence* (pp. 123–137). Norwood, NJ: Ablex.
73. Hild, M., & Spohn, W. (2008). The measurement of ranks and the laws of iterated contraction. *Artificial Intelligence, 172*(10), 1195–1218.
74. Johnson-Laird, P. (2006). *How we reason*. Oxford: Oxford University Press.
75. Katsuno, H., & Mendelzon, A. (1992). On the difference between updating a knowledge base and revising it. In P. Gärdenfors (Ed.), *Belief revision* (pp. 183–203). Cambridge: Cambridge University Press.

76. Kifer, M., Lausen, G., & Wu, J. (1995). Logical foundations of object-oriented and frame-based languages. *Journal of the ACM, 42*, 741–843.
77. Kuhn, T. S. (1970). *The structure of scientific revolutions* (2nd ed.). Chicago: University of Chicago Press.
78. Leblanc, H. (1976). *Truth-value semantics*. Amsterdam: North Holland Publishing Company.
79. Leitgeb, H. (2004). *Inference on the low level: An investigation into deduction, nonmonotonic reasoning, and the philosophy of cognition*. Dordrecht: Kluwer.
80. Liu, Y., Lakemeyer, G., & Levesque, H. J. (2004). A logic of limited belief for reasoning with disjunctive information. In *Principles of Knowledge Representation and Reasoning: Proceedings of the Ninth International Conference (KR2004)*, Whistler (pp. 587–597).
81. Löwe, B. (2006). Set theory with and without urelements and categories of interpretations. *Notre Dame Journal of Formal Logic, 47*(1), 83–91.
82. McCarthy, J., & Hayes, P. (1969). Some philosophical problems from the standpoint of artificial intelligence. In B. Meltzer and D. Michie (Eds.), *Machine Intelligence 4* (pp. 463–502). Edinburgh: Edinburgh University Press.
83. Meheus, J., Straßer, C., & Verdée, P. (2016). Which style of reasoning to choose in the face of conflicting information? *Journal of Logic and Computation, 26*(1), 361–380.
84. Merritt, S. M., & Stix, A. (1997). *Migrating from pascal to C++*. New York: Springer.
85. Meyer Viol, W. P. M. (1995). *Instantial logic. An investigation into reasoning with instances*. ILLC Dissertation Series 1995–11, ILLC, Amsterdam.
86. Minker, J. (1994). Overview of disjunctive logic programming. *Annals of Mathematics and Artificial Intelligence, 12*, 1–24.
87. Minsky, M. A. (1975). Framework for Representing Knowledge. In P. Winston (Ed.), *The psychology of computer vision* (pp. 211–277). McGraw-Hill. http://web.media.mit.edu/~minsky/papers/Frames/frames.html
88. Moulines, C. U. (2010). Metatheoretical structuralism: A general program for analyzing science. *Axiomathes, 20*, 255–268.
89. Nebel, B. (1998). How hard is it to revise a belief base? In D. Dubois & H. Prade (Eds.), *Handbook of defeasible reasoning and uncertainty management systems. Belief change* (pp. 77–145). Dordrecht: Kluwer.
90. Neth, H., & Gigerenzer, G. (2015). Heuristics: Tools for an Uncertain World. In R. Scott & S. Kosslyn (Eds.), *Emerging trends in the social and behavioral sciences: An interdisciplinary, searchable, and linkable resource* (pp. 1–18). New York: Wiley Online Library.
91. Niedermeier, R. (2006). *Invitation to fixed-parameter algorithms*. Oxford: Oxford University Press.
92. Poole, D. (1988). A logical framework for default reasoning. *Artificial Intelligence, 36*, 27–47.
93. Priest, G. (2002). Paraconsistent Logic. In D. Gabbay & F. Guenthner (Eds.), *Handbook of philosophical logic* (Vol. 6. 2nd ed., pp. 287–393). Dordrecht: Kluwer Academic.
94. Ramsey, F. P. (1931). Theories. In R. B. Braithwaite (Ed.), *The foundations of mathematics and other logical essays* (pp. 212–236). London: Routledge and Kegan Paul.
95. Rescher, N. (1964). *Hypothetical reasoning* (Studies in logic and the foundations of mathematics). Amsterdam: North-Holland Publishing Company.
96. Robinson, J. A. (1965). A machine-oriented logic based on the resolution principle. *Journal of the ACM, 12*(1), 23–41.
97. Rossi, F., van Beek, P., & Walsh, T. (2008). Constraint programming. In F. V. Harmelen, V. Lifschitz, & B. Porter (Eds.), *Handbook of knowledge representation* (pp. 181–211). Amsterdam: Elsevier.
98. Rott, H. (2001). *Change, choice and inference: A study of belief revision and nonmonotonic reasoning*. Oxford: Oxford University Press.
99. Samuels, R. (1998). Evolutionary psychology and the massive modularity hypothesis. *British Journal for the Philosophy of Science, 49*(4), 575–602.
100. Samuels, R. (2012). Massive modularity. In E. Margolis, R. Samuels, & S. Stich (Eds.), *Oxford handbook of philosophy of cognitive science* (pp. 60–91). Oxford University Press.

101. Schneider, S. (2011). *Language of thought: A new philosophical direction*. Cambridge, MA: MIT Press.
102. Schöning, U. (2008). *Logic for computer scientists*. Boston: Birkhäuser.
103. Schurz, G. (2011). Abductive belief revision in science. In E. J. Olsson & S. Enqvist (Eds.), *Belief revision meets philosophy of science* (Logic, epistemology, and the unity of science, Vol. 21, pp. 77–104). Dordrecht: Springer.
104. Sneed, J. (1979). *The logical structure of mathematical physics* (2nd ed.). Dordrecht: D. Reidel Publishing Company.
105. Spohn, W. (1988). Ordinal conditional functions: A dynamic theory of epistemic states. In *Causation in decision, belief change, and statistics II* (pp. 105–134). Dordrecht: Kluwer.
106. Stegmüller, W. (1976). *The structure and dynamics of theories*. Berlin/Heidelberg/New York: Springer.
107. Stenning, K., & van Lambalgen, M. (2008). *Human reasoning and cognitive science*. Boston: MIT Press.
108. Stillman, J. (1990). It's not my default: The complexity of membership problems in restricted propositional default logics. In *Proceedings of the 8-th National Conference on Artificial Intelligence (AAAZ-901)* (pp. 571–578).
109. Suppes, P. (1957). *Introduction to logic*. Princeton: D. Van Nostrand Company.
110. Suppes, P. (2002). *Representation and invariance of scientific structures*. Stanford: CSLI Publications.
111. Tseitin, G. S. (1983). On the complexity of derivation in propositional calculus. In J. H. Siekmann & G. Wrightson (Eds.), *Automation of Reasoning: 2: Classical Papers on Computational Logic 1967–1970* (pp. 466–483). Berlin/Heidelberg: Springer.
112. van Benthem, J. (2008). Logic and reasoning: Do the facts matter? *Studia Logica, 88*, 67–84.
113. van Ditmarsch, H., van der Hoek, W., & Kooi, B. (2008). *Dynamic epistemic logic*. Heidelberg: Springer.
114. Verdée, P. (2013). Non-monotonic set theory as a pragmatic foundation of mathematics. *Foundations of Science, 18*(4), 655–680.
115. Wassermann, R. (1999). Resource-bounded belief revision. *Erkenntnis, 50*, 429–446.
116. Weyl, H. (1968). Wissenschaft als symbolische Konstruktion des Menschen. In *Gesammelte Abhandlungen* (Bd. IV, pp. 289–345). Heidelberg/Berlin: Springer.
117. Williams, M.-A. (1994). On the logic of theory base change. In C. MacNish, D. Pearce, & L. M. Pereira (Eds.), *Logics in artificial intelligence* (pp. 86–105). Berlin/Heidelberg: Springer.
118. Winslett, M. S. (1990). *Updating logical data bases*. Cambridge: Cambridge University Press.
119. Wittgenstein, L. (2014). *Tractatus logico-philosophicus: German and English*. Routledge.
120. Woods, J. (2013). *Errors of reasoning: Naturalizing the logic of inference*. London: College Publications.
121. Zach, R. (2015). Lectures on the epsilon calculus http://arxiv.org/abs/1411.3629.
122. Zach, R. (2019). *Sets, logic, computation*. OpenLogicProject. http://builds.openlogicproject.org/courses/phil379/.

Index of Symbols

H. Andreas, *Dynamic Tractable Reasoning*, Synthese Library 420,
https://doi.org/10.1007/978-3-030-36233-1

Author Index

H. Andreas, *Dynamic Tractable Reasoning*, Synthese Library 420,
https://doi.org/10.1007/978-3-030-36233-1

Subject Index

H. Andreas, *Dynamic Tractable Reasoning*, Synthese Library 420,
https://doi.org/10.1007/978-3-030-36233-1

Kolibri 360 GmbH
Ferdinand-Jühlke-Straße 7
99095 Erfurt, Deutschland
produktsicherheit@kolibri360.de